Translation Surfaces

GRADUATE STUDIES
IN MATHEMATICS **242**

Translation Surfaces

Jayadev S. Athreya
Howard Masur

AMERICAN
MATHEMATICAL
SOCIETY
Providence, Rhode Island

EDITORIAL COMMITTEE
Matthew Baker
Marco Gualtieri
Gigliola Staffilani (Chair)
Jeff A. Viaclovsky
Rachel Ward

2020 Mathematics Subject Classification. Primary 32G15, 30F60, 37A10, 37A25, 37-02.

For additional information and updates on this book, visit
www.ams.org/bookpages/gsm-242

Library of Congress Cataloging-in-Publication Data
Names: Athreya, Jayadev S., author. | Masur, Howard, 1949- author.
Title: Translation surfaces / Jayadev S. Athreya and Howard Masur.
Description: First edition. | Providence, Rhode Island : American Mathematical Society, [2024] | Series: Graduate studies in mathematics, 1065-7339 ; volume 242 | Includes bibliographical references and index.
Identifiers: LCCN 2023053390 | ISBN 9781470476779 (paperback) | ISBN 9781470476557 (hardcover) | 9781470476762 (ebook)
Subjects: LCSH: Riemann surfaces. | Ergodic theory. | Dynamics. | AMS: Several complex variables and analytic spaces – Deformations of analytic structures – Moduli of Riemann surfaces, Teichmüller theory. | Functions of a complex variable – Riemann surfaces – Teichmüller theory. | Dynamical systems and ergodic theory – Ergodic theory – One-parameter continuous families of measure-preserving transformations. | Dynamical systems and ergodic theory – Ergodic theory – Ergodicity, mixing, rates of mixing. | Dynamical systems and ergodic theory – Research exposition (monographs, survey articles).
Classification: LCC QA333 .A84 2024 | DDC 515/.93–dc23/eng/20231226
LC record available at https://lccn.loc.gov/2023053390

Copying and reprinting. Individual readers of this publication, and nonprofit libraries acting for them, are permitted to make fair use of the material, such as to copy select pages for use in teaching or research. Permission is granted to quote brief passages from this publication in reviews, provided the customary acknowledgment of the source is given.

Republication, systematic copying, or multiple reproduction of any material in this publication is permitted only under license from the American Mathematical Society. Requests for permission to reuse portions of AMS publication content are handled by the Copyright Clearance Center. For more information, please visit **www.ams.org/publications/pubpermissions**.

Send requests for translation rights and licensed reprints to **reprint-permission@ams.org**.

© 2024 by the authors. All rights reserved.
Printed in the United States of America.

∞ The paper used in this book is acid-free and falls within the guidelines established to ensure permanence and durability.
Visit the AMS home page at **https://www.ams.org/**

10 9 8 7 6 5 4 3 2 1   29 28 27 26 25 24

Contents

Preface		vii
Chapter 1.	Introduction	1
§1.1.	The square torus	1
§1.2.	The space of tori	2
§1.3.	Dynamics	4
§1.4.	Counting	6
Chapter 2.	Three Definitions	9
§2.1.	Polygons	9
§2.2.	Geometric structures	18
§2.3.	Holomorphic 1-forms	21
§2.4.	Saddle connections and cylinders	26
§2.5.	Equivalences	29
§2.6.	Quadratic differentials	33
Chapter 3.	Moduli Spaces of Translation Surfaces	39
§3.1.	Teichmüller and moduli spaces	39
§3.2.	Strata of translation surfaces	44
§3.3.	Coordinates on components of strata	45
§3.4.	Measures	46
§3.5.	Components of strata	50
§3.6.	Delaunay triangulations	52
§3.7.	Finiteness of the MSV measure	59

Chapter 4.	Dynamical Systems and Ergodic Theory	63
§4.1.	Topological dynamics	63
§4.2.	Ergodic theory	67
§4.3.	Low genus examples	70
§4.4.	Number of ergodic measures	81
§4.5.	Interval exchange transformations	86
Chapter 5.	Renormalization	95
§5.1.	A criterion for unique ergodicity	95
§5.2.	Ergodicity of the Teichmüller flow	100
§5.3.	Geodesics, horocycles, and mixing	106
§5.4.	Quantitative renormalization and ergodicity	109
§5.5.	Nonunique ergodicity and Hausdorff dimension	119
Chapter 6.	Counting and Equidistribution	123
§6.1.	Lattice point counting	123
§6.2.	Saddle connections and holonomy vectors	127
§6.3.	Siegel-Veech transforms	128
§6.4.	From counting to dynamics	129
§6.5.	Equidistribution of circles	132
§6.6.	Further counting results	136
§6.7.	Counting and volumes	137
Chapter 7.	Lattice Surfaces	141
§7.1.	Affine diffeomorphisms and Veech groups	142
§7.2.	Examples	144
§7.3.	Cusps, cylinders, and shears	150
§7.4.	Optimal dynamics	151
§7.5.	Homogeneous dynamics	154
§7.6.	Characterizations of lattice surfaces	156
§7.7.	Classifying lattice surfaces	158
§7.8.	Square-tiled surfaces	160
Conclusion		165
Bibliography		167
Index		177

Preface

This book is intended for a second- or third-year graduate topics course on the subject of *translation surfaces* and their moduli spaces, with a focus on associated dynamical systems, counting problems, and group actions.

In recent years, translation surfaces and their moduli spaces have been the objects of extensive study and interest, with connections to widely varied fields including (but not limited to) geometry and topology; Teichmüller theory; low-dimensional dynamical systems; homogeneous dynamics and Diophantine approximation; and algebraic and complex geometry. Contributions to the field have been recognized by major awards, invited addresses at international conferences, and multiple excellent survey articles on various aspects of the subject. However, there has not yet been a textbook written about translation surfaces and the various viewpoints that have been developed that a student or researcher who is interested in the field can use as an accessible introduction.

This book aims to fill that gap, with attention to definitions and an introduction to some of the big ideas in the field, centered on the ergodic properties of translation flows and counting problems for saddle connections, and associated renormalization techniques, without attempting to reach the full state of the art (an aim that is in any case impossible given the speed at which the field is evolving).

In particular, we do not discuss in great detail the subject of the Kontsevich-Zorich cocycle and its Lyapunov exponents and the subject of Rauzy induction, instead giving a brief survey of some of these ideas and mostly referring the reader to the detailed surveys by Forni [70], Zorich [177], Forni-Matheus [71], and Yoccoz [174]. We only scratch the surface, no pun intended, of the subject of *square-tiled surfaces*, which have inspired a lot of interesting connections to arithmetic and combinatorics. We do not discuss in any detail the deep connections to algebraic geometry that have driven some of the exciting recent developments in the field, referring the reader to the survey of Bud-Chen [30].

So what *do* we cover in this book? As a kind of overture, we discuss in Chapter 1 the important motivating example of the flat torus, exploring its geometry, and its associated dynamical and counting problems. The linear flow on the torus and its associated first return map, a rotation of a circle, are amongst the first dynamical systems ever studied. The counting of closed orbits is intricately tied to number theory. We discuss, as motivation, the moduli space of translation surfaces on a torus, a bundle over the well-known modular curve and the action of $GL^+(2,\mathbb{R})$ on this space of translation surfaces.

Translation surfaces are higher-genus generalizations of flat tori. In Chapter 2, we define translation surfaces from three perspectives (Euclidean geometry, complex analysis, and geometric structures) and prove the equivalence of these definitions, something that is often hard to find in the literature. We define when two translation surfaces are equivalent, leading to the definition of moduli spaces of translation surfaces. We show how some translation surfaces arise from *unfolding* billiards in rational polygons. We introduce the subject of half-translation surfaces, or quadratic differentials.

In Chapter 3, we begin with a short introduction to Teichmüller theory and its relation to the study of translation surfaces. We then define moduli spaces and strata of translation surfaces and introduce the $GL^+(2,\mathbb{R})$-action on strata of translation surfaces. We introduce natural period coordinates on strata and use these coordinates to construct the canonical $SL(2,\mathbb{R})$-invariant measure on strata of area 1 translation surfaces. We show the existence of *Delaunay triangulations*, following the work of Masur-Smillie [121]. Understanding these triangulations allows us to show that the measure of strata is finite.

In Chapter 4, we discuss the natural dynamical systems associated to translation surfaces, namely, *linear flows* and their first return maps, *interval exchange transformations*. We explore their ergodicity and mixing properties. Along the way we study an important example of a translation surface flow for which every orbit is dense but not every orbit is equidistributed with respect to Lebesgue measure, a phenomenon that does not occur in the case of linear flows on the torus, where a theorem of Weyl implies that dense orbits are always equidistributed.

In Chapter 5, we show how information about the recurrence properties of an orbit of a translation surface under the positive diagonal subgroup of $SL(2,\mathbb{R})$ (the *Teichmüller geodesic flow*) can be used to get information about the ergodic properties of the associated linear flow on an individual translation surface. This is an example of a major theme in the subject, the interplay between properties of the orbit of a translation surface under the $SL(2, \mathbb{R})$-action and properties of the translation surface itself. We show (§5.2) that the $SL(2, \mathbb{R})$-action on each stratum is ergodic. At the end of Chapter 5, in §5.4, we discuss quantitative versions of renormalization ideas, including our brief discussion of Lyapunov exponents.

As another example of the strength of renormalization ideas, we show in Chapter 6 how the *ergodic properties* of the $SL(2, \mathbb{R})$-action can be used to obtain *counting results* for *saddle connections* and, subsequently, to compute the intrinsic volumes of strata. Finally, in Chapter 7, we discuss examples, characterizations, and properties of surfaces with large affine symmetry groups, known as *lattice* or *Veech* surfaces.

Broadly speaking, the first four and a half chapters of the book are quite detailed and relatively self-contained, while from the middle of Chapter 5 we attempt to communicate some of the big recent ideas in the field, with somewhat less detail, but hopefully a sufficient set of precise references and suggestions for further reading.

Intended uses. This book is intended to be used for a one-semester graduate course introducing students to the subject of translation surfaces. As such, we have included exercises throughout. Attempting these exercises is a crucial part of following along with this book. This book can also be used for researchers interested in getting a big-picture overview of the subject together with concrete details of important definitions and concepts.

Prerequisites and suggested reading. We have attempted to keep this book as self-contained as possible. Recommended prerequisites include first-year graduate courses in complex analysis (at the level of, for example, Conway [38]), measure theory (Royden [142]), and manifolds (do Carmo [47]). Ideally, the reader will have some knowledge of Riemann surfaces and ergodic theory, at the level of, for example, Jost [95] for Riemann surfaces and Walters [170] or Einsiedler-Ward [52] for ergodic theory. A couple of nice books targeted at undergraduates that may be useful to keep alongside our book are Schwartz's *Mostly Surfaces* [150], which introduces the basics of translation surfaces and covers the topology of surfaces in a very accessible fashion, and Davis's [42] recent book *Billiards, Surfaces and Geometry: A Problem-Centered Approach*.

Black Boxes. In order to keep this book relatively self-contained, at several junctures we have to assume certain results whose proofs would be beyond the scope of our book. We have tried to be as clear as possible in indicating we are doing this, by labeling such results using the header *Black Box*.

Black Box. *A Black Box is a result we require but whose proof is beyond the scope of the book.*

For example, we will treat the Riemann-Roch (Black Box 2.3.2) and Gauss-Bonnet (Black Box 2.2.1) theorems as Black Boxes.

Acknowledgements. All mathematicians are products of networks of mentors, peers, and students. Though it is impossible to name everyone from whom we have gained knowledge and perspective, we would like to thank in particular (in alphabetical order, as all good mathematicians practice) Tarik Aougab, Pierre Arnoux, Jon Chaika, Benson Farb, Simion Filip, Giovanni Forni, Anish Ghosh, Chris Hoffman, Pascal Hubert, Chris Leininger, Samuel Lelièvre, Doug Lind, Marissa Loving, Brian Marcus, Jens Marklof, Carlos Matheus Santos, Curtis McMullen, Maryam Mirzakhani, Yair Minsky, Mahan MJ, Martin Möller, Thierry Monteil, Amos Nevo, Priyam Patel, Steffen Rohde, John Smillie, Ferran Valdez, William Veech, Barak Weiss, Gabriela Weitze-Schmithüsen, Amie Wilkinson, Alex Wright, and Anton Zorich for discussions, collaborations, perspectives, and feedback. The first author would particularly like to thank Alex Eskin and Grigory Margulis for their extraordinary mentorship and for introducing him to this circle of ideas. The second author would like to thank Corinna Ulcigrai for her encouragement to write this book. We would like to thank Albert Artiles Calix, Samantha Fairchild, Paige Helms, Anthony Sanchez, and Joshua Southerland for reviewing an early draft of this manuscript. We are deeply grateful to the anonymous reviewers for their careful reading and feedback, which have greatly improved the exposition of the book. We are in particular thankful to Giovanni Forni for his help in improving the exposition of Chapter 5. We are grateful as well to the editorial board of the Graduate Studies in Mathematics Series and in particular to our editor Loretta Bartolini. The authors would also like to thank Arlene O'Sean, the AMS copyeditor, for her excellent job in editing the manuscript. We are of course responsible for any remaining errors in the book. Please contact jathreya@uw.edu if you find any mistakes/typos in the book.

Institutional support. The first author thanks the University of Washington, the University of Washington Royalty Research Fund, the Victor Klee fund, the Pacific Institute for the Mathematical Sciences (PIMS), and the National Science Foundation (via NSF CAREER grant DMS 1559860 and NSF grant DMS 2003528) for financial and institutional support. This work was concluded during his term from July–December 2023 as the Chaire Jean Morlet at the Centre International de Recontres Mathématique (CIRM)-Luminy.

Both authors would like to thank the CIRM, the Park City Mathematics Institue (PCMI), and the Mathematical Sciences Research Institute (MSRI) where parts of this work were completed. The first author would in particular like to thank Kristine Bauer and Ozgur Yilmaz at PIMS and Rafe Mazzeo at PCMI.

The second author thanks the University of Chicago for its support during the time this book was written.

Personal thanks (from the first author). Mathematicians do not live by mathematics alone. I have been extraordinarily lucky to have the support of my family and friends. I would like to thank my father, the late Krishna Balasundaram Athreya, and my mother, Krishna (Rani) Siddhanta Athreya, for their encouragement to pursue mathematics; my siblings Kartik, Avanti, and Ambika for their love, companionship, and humor; and my friends Kemi Adeyemi, Judy Cochran, Jean Dennison, Maria Elena Garcia, Sunila Kale, Jose Antonio Lucero, Rohit Naimpally, Christian Novetzke, Josh Reid, Chandan Reddy, Russell Voth, and Megan Ybarra for their love and support. A special thanks goes to my dear friends Piyali Bhattacharya and Tariq Thachil. I couldn't possibly thank them enough, for everything. Most of all, I thank my partner Radhika Govindrajan for constant emotional and intellectual support.

Personal thanks (from the second author). I am grateful for the support of a loving family and in particular my wife, Elise Masur, who all these years has supported and encouraged my love of mathematics.

Chapter 1

Introduction

1.1. The square torus

A common first example of a compact manifold is the square torus, $\mathbb{C}/\mathbb{Z}[i]$, where $\mathbb{Z}[i] = \{m + ni : m, n \in \mathbb{Z}\}$ is the set of Gaussian integers. This surface has many different interpretations, each of which lends itself to generalization:

Polygon. First, it can be viewed as the unit square

$$\{z = x + iy \in \mathbb{C} : 0 \leq x, y \leq 1\}$$

with parallel sides identified by translation. Precisely, the vertical sides $\{iy : 0 \leq y \leq 1\}$ and $\{1 + iy : 0 \leq y \leq 1\}$ are identified by the translation $z \mapsto z + 1$ and the horizontal sides $\{x : 0 \leq x \leq 1\}$ and $\{i + x : 0 \leq x \leq 1\}$ are identified by the translation $z \mapsto z + i$. See Figure 1.1.

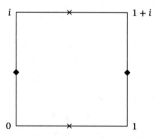

Figure 1.1. A square torus, with side identifications.

Holomorphic 1-forms. The square torus such as $\mathbb{C}/\mathbb{Z}[i]$ has the structure of a *Riemann surface*, that is, a surface with a complex structure, since $\mathbb{Z}[i]$ acts on $\mathbb{C}$ by translations, which are holomorphic. Moreover, the family of holomorphic 1-forms $\{e^{i\theta} dz : \theta \in [0, 2\pi)\}$ on $\mathbb{C}$ descends to $\mathbb{C}/\mathbb{Z}[i]$, since $d(z + c) = dz$.

Constant curvature geometric structure. Since translations are Euclidean isometries, dz induces a (flat) Euclidean metric on $\mathbb{C}/\mathbb{Z}[i]$, inherited from the Euclidean metric on $\mathbb{C}$. In particular, this metric has *constant* (zero) Gaussian curvature.

Dynamical systems. In addition to these three interpretations of the square torus, we have a 1-parameter family of dynamical systems, indexed by $S^1 = [0, 2\pi)/(0 \sim 2\pi)$: given $\theta \in S^1$, we have the flow

$$\phi_t^\theta(z) = (z + te^{i\theta}) \mod \mathbb{Z}[i].$$

Generalizations. When we study general translation surfaces, we will have each of these three interpretations, appropriately modified: polygons, with parallel sides identified by translation; Riemann surfaces, together with holomorphic 1-forms; and flat metrics (with singularities), as well as a 1-parameter family of linear flows. Moreover, we will consider *spaces* of translation surfaces—note that in the above constructions, we never really used the fact that the square torus is a square, but rather that we had a parallelogram with parallel sides identified by translation.

1.2. The space of tori

Families of tori. Given any $v, w \in \mathbb{C}^*$ that are not $\mathbb{R}$-collinear, that is, $v/w \notin \mathbb{R}$, we can construct the lattice $\Lambda = \Lambda(v, w) = \{mv + nw : m, n \in \mathbb{Z}\} \subset \mathbb{C}$ and the associated torus $\mathbb{C}/\Lambda$, which can be viewed as the parallelogram defined by v, w with pairs of sides identified by translation. One choice that corresponds to the square torus is $v = 1, w = i$, so $\Lambda(v, w) = \mathbb{Z}[i]$. Note that any other choice of v, w so that $\Lambda(v, w) = \mathbb{Z}[i]$ would yield the same torus, presented in a different way.

1-forms and flows. Again the family of 1-forms $\{e^{i\theta} dz : \theta \in [0, 2\pi)\}$ descends to $\mathbb{C}/\Lambda$, and again, there is an associated family of linear flows ϕ_t^θ as above. There is a natural *parameter space* of 1-forms living on the space of tori. We note that each 1-form gives a choice of *vertical direction* on the torus. The vertical direction of the 1-form $e^{i\theta} dz$ on the torus $\mathbb{C}/\Lambda$ can be thought of as the set of tangent vectors v to $\mathbb{C}$ such that $e^{i\theta} dz(v)$ is of the form ci where $c > 0$. The 1-form $e^{i\theta} dz$ on $\mathbb{C}/\Lambda$ can be thought of as the 1-form dz on the torus $\mathbb{C}/e^{-i\theta}\Lambda$. We will denote the space of pairs of tori together with 1-forms by $\Omega(\emptyset)$, where $\emptyset$ refers to the fact that 1-form dz does not have any zeros. That is,

$$\Omega(\emptyset) = \{(\mathbb{C}/\Lambda, dz) : \Lambda \text{ a lattice in } \mathbb{C}\}.$$

We emphasize that there will be many v, w for each Λ so that $\Lambda = \Lambda(v, w)$. For example,

$$\Lambda(v, w) = \Lambda(v + w, w) = \Lambda(v, v + w).$$

The $GL^+(2,\mathbb{R})$-action. There is a natural $\mathbb{R}$-linear action of the group $GL^+(2,\mathbb{R})$ consisting of 2×2 invertible matrices with real entries and positive determinant on $\Omega(\emptyset)$, coming from the left $\mathbb{R}$-linear action of $GL(2,\mathbb{R})$ on $\mathbb{C}$ given by viewing $x + iy$ as the column vector $(x \ \ y)$; that is,

$$(1.2.1) \qquad \begin{pmatrix} a & b \\ c & d \end{pmatrix} \cdot (x + iy) = (ax + by) + i(cx + dy),$$

and
$$g \cdot (\mathbb{C}/\Lambda(v,w)) = \mathbb{C}/\Lambda(g \cdot v, g \cdot w).$$

Note that if $\Lambda(v,w) = \Lambda(v',w')$, then $\Lambda(g \cdot v, g \cdot w) = \Lambda(g \cdot v', g \cdot w')$. This action will generalize to spaces of higher-genus translation surfaces. The embedded action of $\mathbb{R}^+$ by multiplication (corresponding to scalar multiples $tI_2, t \in \mathbb{R}^*$ of the identity matrix I_2) scales the area of tori by t^2. The action of S^1 via rotation matrices

$$(1.2.2) \qquad r_\theta = \begin{pmatrix} \cos\theta & -\sin\theta \\ \sin\theta & \cos\theta \end{pmatrix}$$

changes the underlying 1-form by multiplication by $e^{-i\theta}$, or, equivalently, multiplies the lattice Λ by $e^{i\theta}$. Putting these actions together gives a $\mathbb{C}^*$-action on the space $\Omega(\emptyset)$, which can be thought of as preserving the underlying torus but changing the 1-form by scaling; that is, for $\zeta \in \mathbb{C}^*$,

$$\zeta(\mathbb{C}/\Lambda, dz) = (\mathbb{C}/\Lambda, \zeta dz) = (\mathbb{C}/\zeta^{-1}\Lambda, dz).$$

Area 1 tori. As many of the properties we are interested in will be independent of scaling, we will pay special attention to the set of *area 1 tori*—that is, tori for which the parallelogram spanned by v, w has area 1. Note that if $v = x + iy, w = u + is$, the area of the parallelogram (assuming v, w are positively oriented) is given by

$$A(v,w) = \det \begin{pmatrix} x & u \\ y & s \end{pmatrix} = xs - yu$$

$$= \operatorname{Im}(\overline{v}w) = \frac{i}{2}(v\overline{w} - \overline{v}w).$$

The subset of unit-area tori

$$\mathcal{H} = \{(\mathbb{C}/\Lambda(v,w), dz) : A(v,w) = 1\} \subset \Omega(\emptyset)$$

is preserved by the action of the subgroup $SL(2,\mathbb{R})$ of $GL^+(2,\mathbb{R})$ consisting of determinant 1 matrices. The next exercise uses this action to help us to identify the space $\mathcal{H}$ with a quotient of $SL(2,\mathbb{R})$.

Exercise 1.1. *Show that any lattice $\Lambda = \Lambda(v,w)$ with $A(v,w) = 1$ can be written as $\Lambda = g \cdot \mathbb{Z}[i]$, $g \in SL(2,\mathbb{R})$, and if $h \in SL(2,\mathbb{Z}), h \cdot \mathbb{Z}[i] = \mathbb{Z}[i]$. Conclude that the $\mathbb{R}$-linear action of $SL(2,\mathbb{R})$ acts transitively on the space of pairs $(\mathbb{C}/\Lambda(v,w), dz) \in \mathcal{H}$ and the stabilizer of the pair $(\mathbb{C}/\mathbb{Z}[i], dz)$ is the subgroup*

$SL(2, \mathbb{Z})$ *of integer determinant 1 matrices, and thus, by the orbit-stabilzer theorem (see, for example, Artin [5, Proposition 6.4]), we can identify*

$$\mathcal{H} = SL(2, \mathbb{R})/SL(2, \mathbb{Z})$$

via

$$gSL(2, \mathbb{Z}) \mapsto (\mathbb{C}/g \cdot \mathbb{Z}[i], dz).$$

1.3. Dynamics

Returning to the square torus, the dynamics of the linear flow ϕ_t^θ satisfies a natural dichotomy, depending on the angle θ. If $\tan(\theta)$ (that is, the slope of the line at angle θ with the horizontal) is rational, every orbit is periodic. On the other hand, if $\tan(\theta)$ is irrational, every orbit is dense and, moreover, *equidistributed*: if $A \subset \mathbb{C}/\mathbb{Z}[i]$ is an open set, then for all x,

$$\lim_{T \to \infty} \frac{1}{T} |\{0 \le t \le T : \phi_t^\theta(x) \in A\}| = m(A),$$

where m denotes Lebesgue measure on the torus. That is, the proportion of time each orbit spends in a given subset of the torus tends to the measure of the subset. We now show how to prove a discretized version of this using the idea of a *first return map*, which we will explore further in the context of translation surfaces and interval exchange maps in Chapter 4. We identify $I = [0,1]/0 \sim 1 = \mathbb{R}/\mathbb{Z}$ with the unit circle $S^1 = \{z \in \mathbb{C} : |z| = 1\}$ via the map

$$x \mapsto e^{2\pi i x}.$$

Consider lines that make an angle $0 \le \psi \le \pi/2$ with the horizontal. Note that the horizontal lines ($\psi = 0$) on the square torus are all closed. In the discussion below, we will use $[z]$ to denote the coset $z + \mathbb{Z}[i]$. For $\psi \ne 0$, starting at a point $[x_0] \in \mathbb{C}/\mathbb{Z}[i], x_0 \in [0,1)$, define T_ψ to be the *first return map* of the flow ϕ_t^ψ in direction ψ to the closed curve $S = \{[x] : 0 \le x \le 1\}$ (note that $[0] = [1]$), sitting in our square torus $\mathbb{C}/\mathbb{Z}[i]$. The flow trajectory $\phi_t^\psi([x_0])$ returns to S when it hits the line $y = 1$. See Figure 1.2.

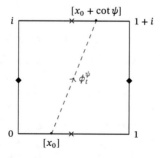

Figure 1.2. The first return map of the flow ϕ_t^ψ to the closed curve S.

1.3. Dynamics

Since
$$\phi_t^\psi([x_0]) = [x_0 + te^{i\psi}] = [(x_0 + t\cos\psi) + it\sin\psi],$$
this return happens at $t = \csc\psi$, and we see that the point of intersection of the line in direction ψ with the copy of S given by $\{[x+i] : 0 \le x \le 1\}$ has x-coordinate $x + \cot(\psi)$, which says that
$$T_\psi([x_0]) = [(x_0 + \cot\psi) \bmod(1)].$$
Via the identification of the interval $[0,1)$ with the circle, the map T_ψ transforms to the map we denote $R_\alpha : S^1 \to S^1$
$$R_\alpha(e^{2\pi i\phi}) = e^{2\pi i(\phi + \alpha)}$$
where $\alpha = \cot\psi$. Depending on whether α is rational or irrational, the map R_α has very different behavior. We leave as an exercise the proof that if α is rational, the map R_α is periodic.

Exercise 1.2. *Prove that if $\alpha \in \mathbb{Q}$ and written in lowest terms as $\alpha = p/q$ with $p, q \in \mathbb{Z}$, $\gcd(p,q) = 1$, then every orbit of R_α is periodic with period q.*

Equidistribution. For α irrational, we have a classical equidistribution result, due originally to Weyl [171]. We say a sequence of points $\{x_n\}_{n\ge 1}$ in a topological space X equidistributes with respect to a (probability) measure η if the sequence of probability measures
$$\sigma_N := \frac{1}{N}\sum_{n=1}^N \delta_{x_i}$$
converges in the weak-* sense to η; that is, for every continuous function $f \in C(X)$,
$$\lim_{N\to\infty} \int_X f d\sigma_N = \lim_{N\to\infty} \frac{1}{N}\sum_{n=1}^N f(x_i) = \int_X f d\eta.$$

Theorem 1.3.1. *If $\alpha \notin \mathbb{Q}$, then every orbit of R_α is dense and in fact every orbit equidistributes in the circle with respect to the Lebesgue probability measure $d\phi$, where we parameterize $S^1 = \{e^{2\pi i\phi} : 0 \le \phi < 1\}$.*

Proof. Weyl [171] showed (see Exercise 1.3) that a sequence $\{x_n\}_{n\ge 1} \subset [0,1)$ becomes equidistributed with respect to Lebesgue measure if and only if for every positive integer k
$$\lim_{N\to\infty} \frac{1}{N}\sum_{n=1}^N e^{2\pi ikx_n} = 0 = \int_{S^1} e^{2\pi ik\phi} d\phi.$$

We use this result to show equidistribution of every orbit of R_α. Writing $e^{2\pi i x_n} = R_\alpha^n(e^{2\pi i x})$ we have

$$(1.3.1) \quad \frac{1}{N}\sum_{n=1}^{N} e^{2\pi i k x_n} = \frac{1}{N}\sum_{n=1}^{N} e^{2\pi i k(x+n\alpha)} = \frac{1}{N} e^{2\pi i k x} \sum_{n=1}^{N} e^{2\pi i k n \alpha}$$

$$= \frac{e^{2\pi i(kx+k\alpha)}}{N} \frac{1 - e^{2\pi i k N \alpha}}{1 - e^{2\pi i k \alpha}}.$$

Since α is irrational, $1 - e^{2\pi i k \alpha} \neq 0$. Since $|e^{2\pi i t}| = 1$ for any real t, the limit of the above quantity is 0 as $N \to \infty$. $\square$

Exercise 1.3. *Prove Weyl's theorem that a sequence $\{x_n\}_{n\geq 1}$ on S^1 becomes equidistributed if and only if for every positive integer k*

$$\lim_{N\to\infty} \frac{1}{N} \sum_{n=1}^{N} e^{2\pi i k x_n} = 0$$

assuming the Stone-Weierstrass theorem (see, for example, Rudin [143, Theorem 1.26]) which states that the set of trigonometric polynomials

$$\left\{ \sum_{j=1}^{J} a_j f_j : a_j \in \mathbb{C} \right\}$$

consisting of finite linear combinations of functions f_j given by

$$f_j(\theta) = e^{2\pi i j \theta} = \cos j\theta + i \sin j\theta$$

is dense *in the set $C(S^1)$ of complex-valued continuous functions on S^1.*

1.4. Counting

We've seen above (Exercise 1.2) that rational slopes correspond to (families of) *periodic* trajectories for linear flows on the square torus $\mathbb{C}/\mathbb{Z}[i]$. We can then ask to *count* the periodic trajectories, graded by length. To remove ambiguity, we choose one representative from each family, by considering only periodic trajectories through the point 0. These trajectories are in 1-to-1 correspondence with the set P of *primitive* Gaussian integers,

$$P = \{m + ni : m, n \in \mathbb{Z}, \gcd(m, n) = 1\}.$$

Defining

$$N_{\text{prim}}(R) = \#(P \cap B(0, R)),$$

we will argue that

$$\lim_{R \to \infty} \frac{N_{\text{prim}}(R)}{\pi R^2} = \frac{1}{\zeta(2)},$$

1.4. Counting

where $\zeta(s) = \sum_{n=1}^{\infty} n^{-s}$. The constant $1/\zeta(2) = \frac{6}{\pi^2}$ is related to the natural volume of the moduli space of tori. We will see similar quadratic asymptotics and relations to volumes of moduli spaces for general translation surfaces in Chapter 6.

Tiling. Without the primitivity condition, the count
$$N(R) = \#\{m + ni \in \mathbb{Z}[i] : m^2 + n^2 \leq R^2\}$$
is seen by a tiling argument to grow asymptotically like πR^2.

Exercise 1.4. Prove
$$\lim_{R \to \infty} \frac{N(R)}{\pi R^2} = 1.$$

Counting primitive points. To count primitive vectors, first note that
$$\mathbb{Z}[i] \setminus \{0\} = \bigsqcup_{n \in \mathbb{N}} nP,$$
so if
$$N_{\text{prim}}(R) = \#(P \cap B(0, R))$$
was asymptotic to $c_P \pi R^2$ for some c_P, we would have, for $n \in \mathbb{N}$,
$$\#(nP \cap B(0, R)) = \#(P \cap B(0, R/n)) = N_{\text{prim}}(R/n)$$
is asymptotic to $\frac{c_P}{n^2} \pi R^2$ for all $n \in \mathbb{N}$. Therefore,
$$c_P \sum_{n \in \mathbb{N}} n^{-2} = 1;$$
that is, $c_P = \frac{1}{\zeta(2)} = 6/\pi^2$. Alternately, note that a heuristic for the probability that an element $m + ni \in \mathbb{Z}[i]$ is in P is given by the Euler product over all primes p
$$\prod_{p} (1 - p^{-2}) = \frac{1}{\zeta(2)},$$
since the probability that m and n are both divisible by a particular prime p is p^{-2}. We leave formalizing this intuition as a (nontrivial) exercise:

Exercise 1.5. Prove
$$\lim_{R \to \infty} \frac{N_{\text{prim}}(R)}{\pi R^2} = \frac{1}{\zeta(2)} = 6/\pi^2.$$

Orbits. Finally, we note that $\mathbb{Z}_{\text{prim}}[i] = SL(2, \mathbb{Z}) \cdot i$ consists of a *single* orbit of $SL(2, \mathbb{Z})$ acting on $\mathbb{C}^*$, which will be relevant for us in particular in Chapter 7, where we study particular families of translation surfaces which resemble the torus in many important ways.

Chapter 2

Three Definitions

In the introduction, we discussed flat tori from three perspectives: as parallelograms with parallel sides identified, Riemann surfaces with holomorphic 1-forms, and constant curvature metrics. We discussed the dynamics of linear flows on tori and their associated first return maps, a group action on the space of tori, and natural associated counting problems. Translation surfaces are higher-genus generalizations of tori, and once again, there are several different perspectives, each of which allows a different definition. We start in §2.1 with the most natural Euclidean geometric perspective, basic properties, important examples, and the connection to rational polygonal billiards. We move onto a geometric structures perspective in §2.2, and then to the complex analytic perspective in §2.3. The equivalence of these three perspectives is at the heart of the study of geometry and dynamics of translation surfaces and their moduli spaces and is often difficult to find fully explained in the literature, though the excellent Séminaire Bourbaki notes of Quint [**139**] do provide a good level of detail. We prove this equivalence in §2.5. Before that, we introduce the key concepts of saddle connections and their holonomies in §2.4.

2.1. Polygons

We start with a finite collection of polygons $P = \{P_1, \ldots, P_n\}, P_i \subset \mathbb{C}$, in the plane $\mathbb{C}$ such that the collection $\mathcal{S}$ of sides of P is grouped into pairs $(s_1, s_2), s_i \in \mathcal{S}$, such that each side s of P belongs to exactly one pair and the two sides (s_1, s_2) in each pair are parallel and of the same length. To build a surface out of this collection, orient each side so that the Euclidean translation taking one side to the other preserves the orientation. For each pairing, we require that the polygon of one of the oriented sides lies to the left of the oriented side and the polygon corresponding to the other side lies to the right. This guarantees the result is a closed orientable surface.

Identifying vertices. The identification of sides by the translation induces an identification of a collection of vertices of the polygons. The result is a closed surface which we will denote by $S = (X, \omega)$. This notation reflects that since this construction involves gluing with translations, which are holomorphic, it yields a Riemann surface structure on the closed surface punctured at the vertices. Working in charts near the points, we can see that the holomorphic charts will be bounded near these points, that we can extend the holomorphic structure to obtain a Riemann surface structure on the closed surface X by filling in the punctures (see, for example, the discussion of removable singularities in Schlag [**147**, §2.3]). The Riemann surface X comes equipped with a holomorphic 1-form coming from pulling back dz (since $d(z + c) = dz$), which can potentially vanish at the vertices. We distinguish between a collection of polygons P and the rotated copy $e^{i\theta} P$: if the original surface yields $S = (X, \omega)$, the rotated surface will yield the pair $S_\theta = (X, e^{i\theta}\omega)$. That is, these translation surfaces differ since the choice of vertical direction differs by angle θ.

Metrics and flows. Since translations are Euclidean isometries, the surface S inherits a local Euclidean metric from the polygons, except possibly at identified vertices. Since translations preserve lines of a fixed slope and directions on those lines, we can consider the families of Euclidean lines of a given slope on the surface through any point that is not a singularity. For any angle and time t we will be able to flow a point $p \in S$ at angle θ distance t (unless it hits a singularity). As in the introduction, this straight line flow will be denoted ϕ_t^θ.

Cone angles. For any vertex, we can take a point on a side near the vertex and traverse a small circle around the vertex following it by the side identifications until it returns. The angle traversed by the loop is called the *cone angle* at the singularity. We will see below that this is an integer multiple of 2π.

2.1.1. Examples.

The torus. We present several examples. Our first example, which we discussed in Chapter 1 is when P is a single parallelogram, with sides $z, w \in \mathbb{C}$. Identifying opposite sides by translation gives a torus with a Euclidean structure. Equivalently, we have a lattice $\Lambda(z, w) = \{mz + nw : m, n \in \mathbb{Z}\} \subset \mathbb{C}$, generated by the sides of the parallelogram and our resulting surface is $\mathbb{C}/\Lambda$. See Figure 2.1. The vertices identify to a single point. The sum of the interior angles at the vertices of the parallelogram is 2π and this is exactly the total change of angle along a small loop surrounding the vertex on the resulting torus. Thus the torus is Euclidean at the vertex as well.

2.1. Polygons

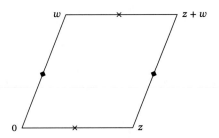

Figure 2.1. A parallelogram torus $\mathbb{C}/\Lambda(z, w)$.

Linear flows. As we discussed in Chapter 1, for a given slope the lines of that slope are either all closed or all dense, and in the latter case they are in fact equidistributed. In the case that the parallelogram is the unit square (for example, $z = 1, w = i$) the resulting lattice is $\mathbb{Z}[i]$; if the slope is rational, the lines are closed, and if the slope is irrational, the lines are dense and equidistributed. Counting the number of periodic families of length at most R corresponds to counting Gaussian integers $m + ni$ with $\gcd(m, n) = 1$ and $m^2 + n^2 \leq R^2$ and this count $N(R)$ grows asymptotically like $\frac{\pi R^2}{\zeta(2)} = \frac{6}{\pi} R^2$.

Hexagons. A next natural example is the regular hexagon. Once again, the resulting surface is a torus: One can easily check that when identifying opposite sides, the six original vertices collapse to two points on the surface, each with a total angle of $3 \times 2\pi/3 = 2\pi$.

Exercise 2.1. *Show how to cut and reglue the regular hexagon into a parallelogram with opposite sides identified. Give explicit z, w so that the hexagonal torus is $\mathbb{C}/\Lambda(z, w)$. Show that the resulting lattice $\Lambda(z, w)$ is the set of Eisenstein integers $\mathbb{Z}[e^{i\pi/3}]$.*

Octagons. We note that squares, hexagons, and parallelograms all tile the plane $\mathbb{C}$ by translation. This is not the case for our next examples. An example of a *genus* 2 surface is given by an octagon whose opposite sides are parallel and of the same length. Identifying these edges yields a surface of genus 2.

Exercise 2.2. *Check that under these edge pairings the eight vertices are identified to a single point and the total angle at this point is 6π.*

Computing genus. There are various ways of seeing that the resulting surface has genus 2.

Euler characteristic. We can directly compute the Euler characteristic of the resulting surface S.

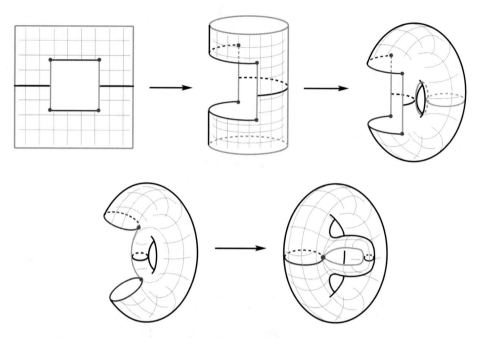

Figure 2.2. A genus 2 translation surface obtained from a square with a smaller square cut out of it, with identifications indicated by the colored sides. The final pink point is the singular point, and it can be seen that 12 squares come together at this point, yielding a cone angle of $12 \times \pi/2 = 6\pi$.

Black Box 2.1.1 ([150, §3.4]). *The* Euler characteristic *of a compact genus g surface Σ_g is $2 - 2g$. Given any polygonal decomposition of Σ_g, the Euler characteristic χ is expressed as*

$$2 - 2g = \chi = f - e + v,$$

where f is the number of faces, e the number of edges, and v the number of vertices in the decomposition.

Decomposing the surface. Viewing the surface as an octagon with opposite sides glued yields a polygonal decomposition, with 1 face, 4 edges, and 1 vertex (following from the exercise), so $f = 1, e = 4, v = 1$. Thus $f - e + v = -2$, so $g = 2$.

Another genus 2 example. Another genus 2 example is illustrated in Figure 2.2. We start with a square with a smaller square deleted and identify edges as indicated in Figure 2.2.

Notation. In notation that we will introduce more fully in Chapter 3, we say the translation surface belongs to the stratum $\Omega(2)$ which means it has a single cone point with cone angle that is $2 \cdot 2\pi$ in excess of 2π. The holomorphic 1-form resulting from pulling back dz has a single zero of order 2 at this point.

2.1. Polygons

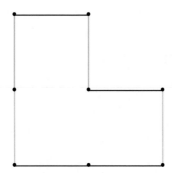

Figure 2.3. An L-shaped translation surface. Identifying parallel sides as indicated by color by translation leads to a genus 2 surface.

An important special case is the regular octagon which we will discuss in more detail later.

Decagons. Another example is a decagon, whose opposite sides have the same length, which are identified by translations to produce a surface.

Exercise 2.3. *Show there are two singularities after side identifications, each has cone angle 4π, and the resulting surface has genus 2.*

Notation. We say that this translation surface belongs to the space $\Omega(1,1)$ which means it has two cone points each with angle in excess of 2π by $1 \cdot 2\pi$. Equivalently the 1-form resulting from pulling back dz has two zeros each of order 1.

$4g$- and $4g+2$-gons. The above examples can be generalized to higher-genus surfaces. We leave the following as an exercise:

Exercise 2.4. *Let $g \geq 1$ be an integer. Starting with a $4g$-gon whose opposite sides are parallel and of the same length, show that identifying opposite sides by translation yields a genus g surface with one cone point with angle $2\pi(2g-1)$. Show that identifying opposite sides of a $4g+2$-gon yields a genus g surface with two cone points, each with angle $2\pi g$. In our notation, we will say the surfaces belong to $\Omega(2g-2)$ and $\Omega(g-1, g-1)$, respectively.*

L-shapes. Our polygons are not required to be convex. As an example, consider the L-shaped polgyon as in Figure 2.3, with side identifications according to the color scheme.

Exercise 2.5. *Show that the resulting surface is a surface in $\Omega(2)$, that is, a surface of genus 2 with one cone point with angle 6π.*

Families of surfaces. Note that in all of our examples, by varying the lengths and directions of the sides, we in fact obtain *families* of such surfaces.

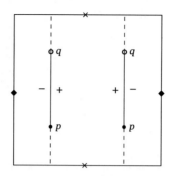

Figure 2.4. A translation surface obtained by identifying opposite sides of the larger square and interior sides with gluings indicated by + and − yields a genus 2 translation surface with two cone points of angle 4π, at the points p and q.

More examples in $\Omega(1,1)$. For another family of examples, take a parallelogram and two oriented parallel line segments within it of the same length. Label each segment with + and −, and identify the + and − sides of each of these segments, and identify opposite sides of the larger parallelogram. In Figure 2.4 the parallel line segments are the solid lines.

Exercise 2.6. *Show that the surface that arises from this construction has genus 2, by showing that it is a torus with two boundary components which are glued to each other. Show further that each of the two endpoints of the segment gives rise to a cone angle 4π singularity so that the surface is in $\Omega(1,1)$.*

Unfolding. In Figure 2.4, the parallelograms are squares and the slits are parallel to the vertical sides. As we vary the side lengths and the lengths of the slits, this family of examples will yield interesting dynamics for linear flows described later in Chapter 4. This family of surfaces will arise as *unfoldings* of a billiard table consisting of a square with a barrier, a procedure we'll describe in general in §2.1.3.

Slit tori. We can take the surface given by Figure 2.4 and realize it differently. Cut the surface along the pair of *dotted* vertical segments that join • and ○. The result is a pair of tori shown in Figure 2.5. (The dotted lines of Figure 2.4 remain dotted lines in this figure.) In Figure 2.5, the two tori are identical, but that is not a requirement of the construction. What is needed in general is a pair of tori and parallel segments on each of the same length which are then glued together. In point of fact it is a theorem of McMullen [**125**, Theorem 1.7] that *every* genus 2 surface with a pair of singularities arises via such a slit construction in infinitely many ways. The particular symmetric case and the fact that there are infinitely many ways of representing it by gluing tori together will be extremely important later in the chapter on dynamical systems.

2.1. Polygons

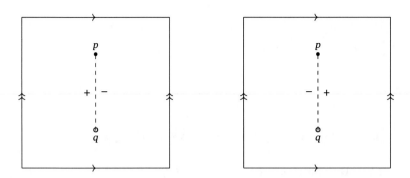

Figure 2.5. A translation surface obtained by gluing two identical square tori along a slit, with each square having opposite sides identified by translation, and the slit gluing indicated by + and −. This yields a genus 2 translation surface with two cone points of angle 4π, at the points p and q.

2.1.2. Properties of translation surfaces. We now record some consequences of this construction, starting with the fact mentioned above that cone angles at the singular points arising from this construction are integer multiples of 2π.

Lemma 2.1.2. *The cone angle at any vertex is an integer multiple of 2π.*

Proof. Given a vertex v and adjacent sides e_1 and e_2, let $\theta_{1,2}$ be the interior angle at v between the edges e_1 and e_2. For $i = 1, 2$ choose $z_i \in e_i$ close to v. Move from z_1 to z_2 in a tiny neighborhood of v sweeping out angle $\theta_{1,2}$. Now e_2 is identified with another edge, still denoted e_2. Again denote by z_2 the identified point. Now there is an edge e_3 such that e_2 and e_3 are adjacent edges at v. These edges subtend an angle $\theta_{2,3}$. Join z_2 to a point $z_3 \in e_3$ across the sector defined by these two edges. The identified e_3 is adjacent at a vertex v to an edge e_4. Since there are only finitely many edges and these are identified in pairs, we must reach an edge e_n, whose identified e_n is adjacent to e_1 at v. The resulting loop sweeps out a total angle

$$\sum_{i=1}^{n} \theta_{i,i+1},$$

where the indices should be viewed mod n; that is, $n + 1 \cong 1$. Since the edges identified as e_1 are parallel, this total angle must be an integer multiple of 2π. □

Cone angles and singularities. Next, we record the relationship between the cone angles of singularities and the genus of the resulting surface.

Theorem 2.1.3. *Let $p_1, \ldots, p_k$ be singularities of ω with cone angle $2\pi(n_i + 1)$ at p_i. Then*

$$\text{(2.1.1)} \qquad \sum_{i=1}^{k} n_i = 2g - 2$$

where g is the genus of the Riemann surface X underlying ω.

Excess angle. The quantity $2\pi n_i = 2\pi(n_i + 1) - 2\pi$ is called the *excess angle* above 2π at the point p_i.

Proof. Triangulate each polygon so that the vertices of the triangles coincide with vertices of the polygon. Let f be the number of triangles. Each edge of a triangle is identified with an edge of another triangle. The number of edges is therefore $\frac{3}{2}f$ and k is the number of vertices. Then

$$\chi(X) = 2 - 2g = f - \frac{3}{2}f + k.$$

Since each triangle has total angle π, the sum of the angles from all triangles is

$$\pi f = (4g - 4)\pi + 2\pi k.$$

Then

$$2\pi \sum_{i=1}^{k} n_i = \pi f - 2\pi k = (4g - 4)\pi.$$

Dividing by 2π, we obtain (2.1.1). □

2.1.3. Billiards and unfolding. Much of the interest in translation surfaces arises from the study of the dynamics of billiards in polygons. Suppose Q is a simply connected polygon in the plane whose interior vertex angles are all rational multiples of π. These will be called *rational billiard tables*. The edges of Q are oriented and labelled.

Exercise 2.7. *Show that the rationality condition on the angles implies that the group $G_Q \subset O_2(\mathbb{R})$ generated by reflections in the lines through the origin parallel to the sides of Q is finite.*

Billiard dynamics. Billiards in a polygon Q refers to the dynamics of a point mass moving at unit speed, with no friction, and elastic collisions with the sides; see Figure 2.6.

2.1. Polygons

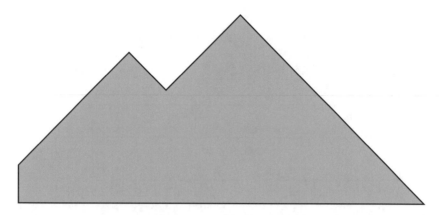

Figure 2.6. A nonconvex rational billiard table.

Phase space. Precisely, billiards on Q are a dynamical system (in this case, a flow) on the phase space $Q \times S^1$, and the finiteness of G_Q implies that this space decomposes into invariant copies of Q—if the initial velocity is $e^{i\theta}$, the set of potential velocities is the G_Q-orbit of θ.

Unfolding. Using this observation, we now show how *unfolding* the polygon gives rise to a translation surface ω_Q and how billiard lines in a direction θ give rise to the straight line flow ϕ_t^θ on ω_Q. This is a construction due to Fox-Kershner [73] (who considered linear flows on polyhedra with rational angles) and independently to Zemljakov-Katok [99] (who considered rational billiards). The key idea is as follows: when a billiard trajectory in direction θ strikes an edge e of Q, rather than reflecting in the side e, we reflect the polygon Q in e and continue the path in a straight line in the same direction θ in the reflected polygon Q'. The polygon Q' again has oriented labeled edges. Now when the billiard path hits an edge of Q' we again reflect Q' in that edge. We continue doing that until we reach a polygon $Q^{(n)}$ which has an edge $\hat{e}$ which is parallel to the labeled edge $\hat{e}$ in a different polygon $Q^{(m)}$, with the same orientation. Because the group G_Q is finite, for every edge $\hat{e}$ we will see the edge again parallel to itself with the same orientation after at most $|G_Q|$ such reflections. We then glue $Q^{(n)}$ to $Q^{(m)}$ along $\hat{e}$ by parallel translation. The result is a translation surface ω_Q, and (unit-speed) billiard paths on the original table become (unit-speed) straight lines on ω_Q.

Billiards in a square. The first example is billiards in a square, which unfolds to a larger polygon, consisting of 4 copies of our original square, with parallel sides with the same orientation identified by translation, resulting in a torus; see Figure 2.7. For example the left vertical edge of the lower left square is glued to the right vertical edge of the lower right adjacent square. The lower horizontal edge of the lower left square is glued to the upper horizontal edge of the upper right square and so forth.

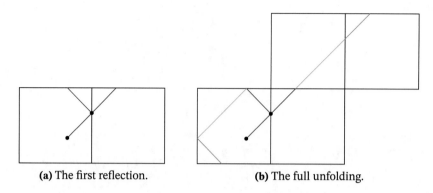

(a) The first reflection. (b) The full unfolding.

Figure 2.7. Unfolding billiards in a square.

A right triangle. Our next example is a (right) triangle with angles $(\pi/8, 3\pi/8, \pi/2)$. Unfolding this triangle yields a regular octagon with parallel sides identified by translation, as seen in Figure 2.8.

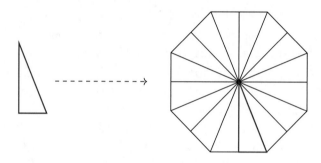

Figure 2.8. Unfolding the $(\pi/8, 3\pi/8, \pi/2)$-triangle to the translation surface given by regular octagon.

Exercise 2.8 (Hard)**.** *Let a, b, c be positive rational numbers with $a + b + c = 1$. Compute the genus and number of singularities of the translation surface that results from unfolding billiards in the triangle with angles $a\pi, b\pi, c\pi$. See, for example, Kenyon-Smillie* **[102]** *for a detailed treatment of billiards in rational triangles.*

Exercise 2.9. *Compute the genus and number of singularities of the translation surface that results from unfolding billiards in the regular n-gon.*

2.2. Geometric structures

We now turn to our second way of defining a translation surface, which generalizes the idea of the torus as a surface with a flat metric and a specified direction. In higher genus, to have a flat metric, we will need to have *singular*

2.2. Geometric structures

points. We start with a closed genus g topological surface M and a finite subset of M denoted Σ. A *translation structure* is given by the following data:

Regular points. On $M \setminus \Sigma$ we have an atlas of coordinate charts U_z and coordinate functions
$$\phi_z : U_z \to \mathbb{C}$$
such that for $U_z \cap U_w \neq \emptyset$, the transition function
$$\phi_z \circ \phi_w^{-1} : \phi_w(U_z \cap U_w) \to \phi_z(U_z \cap U_w)$$
is a *translation*, that is of the form
$$\phi_z \circ \phi_w^{-1}(\zeta) = \zeta + c$$
for a constant $c \in \mathbb{C}$.

No rotational holonomy. We note that rotations are *not* allowed as transition functions. Since translations preserve the Euclidean metric, this allows us to build a local Euclidean metric $|dz|$ on $M \setminus \Sigma$, and we have a well-defined notion of direction (and so can talk about families of parallel lines). For a given direction θ, if we choose a unit vector at each point of $M \setminus \Sigma$ in direction θ, we have a vector field whose flow lines ϕ_t^θ are these parallel lines on the surface. The flow lines for ϕ_t^θ and $\phi_t^{\theta+\pi}$ are the same, just pointing in opposite directions.

Singular points. For $p \in \Sigma$, the structure around p is as follows. For an integer $k \geq 1$ take Euclidean half-planes $\mathbb{H}_j$, $j = 1, \ldots, 2k+2$, bounded by the real axis. For $j = 2m+1$ odd, the half-planes have positive y-coordinate, that is,
$$\mathbb{H}_{2m+1} = \{z_{2m+1} = x_{2m+1} + iy_{2m+1} : y_{2m+1} \geq 0\},$$
and for $j = 2m$ even, the half-planes have negative y-coordinate, that is,
$$\mathbb{H}_{2m} = \{z_{2m} = x_{2m} + iy_{2m} : y_{2m} \leq 0\}.$$
For $1 \leq j \leq 2k+2$ and $j = 2m+1$ odd, glue the negative real axis
$$\{x_j : x_j < 0\}$$
of $\mathbb{H}_j$ to the negative real axis
$$\{x_{j+1} : x_{j+1} < 0\}$$
of $\mathbb{H}_{j+1}$ and for $j = 2m$ even, glue the positive real axis
$$\{x_j : x_j > 0\}$$
of $\mathbb{H}_j$ to the positive real axis
$$\{x_{j+1} : x_{j+1} > 0\}$$
of $\mathbb{H}_{j+1}$. Here we view the indices of the half-planes mod $2k+2$, so $\mathbb{H}_{2k+3} = \mathbb{H}_1$. The half-planes are all glued at 0. The angle traced out by a loop around 0 is

$(2k + 2)\pi$. We have a cone type singularity with cone angle $(2k + 2)\pi$. For $p \in \Sigma$, we define $\eta(p) = 2k\pi$ to be the *excess* angle.

Neighborhood of singular points. Given a neighborhood U_p of a singular point p and a coordinate map ϕ_p from U_p to this collection of glued half-planes and a point $z \in M \setminus \Sigma$ and coordinate chart $\phi_z : U_z \to \mathbb{C}$ such that $U_p \cap U_z \neq \emptyset$, we restrict U_z so that $\phi_p(U_p \cap U_z)$ lies in a single half-plane. We then require that

$$\phi_z \circ \phi_p^{-1} : \phi_p(U_z \cap U_p) \to \phi_z(U_z \cap U_p)$$

is again a *translation*.

Metric in singular neighborhoods. We can define a metric in a neighborhood of $p \in \Sigma$ as follows: choose polar coordinates (r, θ) based at p and define a metric

$$ds^2 = r^k(dr^2 + (rd\theta)^2).$$

The curvature K at p is defined to be

$$K(p) = -2k\pi.$$

Notice that when $k = 0$, then $K(p) = 0$, so the metric

$$ds^2 = dr^2 + r^2 d\theta^2$$

is the Euclidean metric; the cone angle is 2π. We do not consider such points to be singularities. We recall, as a Black Box, the Gauss-Bonnet theorem.

Black Box 2.2.1 (Gauss-Bonnet [**150**, §17]). *Let S be an orientable Riemannian surface S, possibly with boundary. Let K be the Gaussian curvature and k_g the geodesic curvature on ∂S. Then*

$$\int_S K dA + \int_{\partial S} k_g ds = 2\pi \chi(S).$$

Here $\chi(S)$ is the Euler characteristic of the surface and $\int_{\partial S} k_g ds$ is the sum of the corresponding integrals along the smooth part of ∂S plus the sum of the turning angles at the corners of ∂S.

In the case of closed surface

$$\int_S K = 2\pi \chi(S) = 2\pi(2 - 2g).$$

Since $K = 0$ except at points $p \in \Sigma$, the Gauss-Bonnet theorem yields, as in our polygon definition

$$\sum_{p \in \Sigma} K(p) = \sum_{p \in \Sigma} -2k\pi = 2\pi(2 - 2g).$$

We should think of the singular points as being δ-masses of curvature; that is, all the negative curvature of the surface is concentrated at this finite set of points. A precise *combinatorial* version of Black Box 2.2.1 (see [**150**, Theorem

2.3. Holomorphic 1-forms

17.1] states that if S is a closed cone surface (that is, flat except for a finite set of cone points Σ), we have

(2.2.1) $$\sum_{p \in \Sigma} K(p) = 2\pi \chi(S).$$

2.3. Holomorphic 1-forms

For our third definition, recall that our construction of the torus $\mathbb{C}/\Lambda$, for a lattice $\Lambda \subset \mathbb{C}$, yielded in addition to a Riemann surface structure (since the action of Λ on $\mathbb{C}$ by translations is holomorphic) a holomorphic 1-form dz, coming from the 1-form dz on $\mathbb{C}$ and the fact that $d(z + c) = dz$ for any $c \in \mathbb{C}$. This gives us the motivation for the classical complex analytic definition of translation surfaces. Let X be a compact Riemann surface of genus $g \geq 1$. That is, there is an atlas of charts (open sets $\{U_\alpha\}_{\alpha \in A}$ together with homeomorphisms $z_\alpha : U_\alpha \to \mathbb{C}$) such that the transition maps

$$z_\alpha \circ z_\beta^{-1} : z_\beta(U_\alpha \cap U_\beta) \to z_\alpha(U_\alpha \cap U_\beta)$$

are biholomorphic. A *holomorphic 1-form* or *Abelian differential* ω on X assigns to each local coordinate z_α a holomorphic function $\phi_{z_\alpha}(z_\alpha)$ such that in the overlapping charts

(2.3.1) $$\phi_{z_\beta}(z_\beta)\frac{dz_\beta}{dz_\alpha} = \phi_{z_\alpha}(z_\alpha).$$

In the language of sections, a holomorphic 1-form ω is a section of the holomorphic cotangent bundle of X, known as the *canonical bundle*. Given a constant $t \in \mathbb{C}$ we can scale any holomorphic 1-form ω by t, and we call this 1-form $t\omega$. Similarly, we can add two holomorphic 1-forms ω_1, ω_2 on X, to form $\omega_1 + \omega_2$. Note that 0 is a well-defined holomorphic 1-form. Thus, the set of holomorphic 1-forms on a fixed Riemann surface X forms a vector space $\Omega(X)$.Holomorphic 1-form|)

Notation. We will use for the rest of the book the notation ω to denote a translation surface, with the underlying Riemann surface denoted by X only mentioned when needed. The value of a holomorphic 1-form ω at a point $p \in X$ is not well-defined, since it depends on the choice of coordinate chart and therefore the function ϕ. However, since $\frac{dz_\beta}{dz_\alpha}(p) \neq 0$, in overlapping charts, the statement that ϕ_{z_α} is nonzero or has a zero of order k at p is well-defined, independent of charts. We say then that ω has a zero of order k at p. Let Σ denote the set of zeros of ω.

Classical results. We recall two classical results in Riemann surface theory; see, for example, [95]. Fixing our compact genus g Riemann surface X, we note that, following from the Gauss-Bonnet theorem (Black Box 2.2.1) for

example, that for $\omega \in \Omega(X)$ with zeros of order k_i, $i = 1, \ldots, n$, that
$$\sum_{i=1}^{n} k_i = 2g - 2.$$
We record the following:

Black Box 2.3.1 ([95, §5.5]). *$\Omega(X)$ is a complex vector space of dimension g.*

Divisors and zeros. Given a Riemann surface X, let $\text{Div}(X)$ be the free Abelian group generated by the points of X, let $K(X)$ be the field of meromorphic functions on X, and let $M(X)$ be the vector space of meromorphic 1-forms on X. An element $D \in \text{Div}(X)$ has the form
$$D = \sum_{P \in X} n_P P, \quad n_P \in \mathbb{Z}$$
where $n_P = 0$ for all but finitely many $P \in X$. The *degree* of D is given by
$$\deg(D) = \sum_{P \in X} n_P \in \mathbb{Z},$$
and the map $\deg : \text{Div}(X) \to \mathbb{Z}$ is a group homomorphism. Associated to any meromorphic function $f \in K(X)$ on X is the divisor (f) which records the orders of the zeros and poles of f. Similarly, associated to any meromorphic 1-form $\eta \in M(X)$, we can associate the divisor (η). We can compare divisors via a partial order: if
$$D_1 = \sum_{P \in X} n_P P, \quad D_2 = \sum_{P \in X} m_P P,$$
we say $D_1 \leq D_2$ if $n_P \leq m_P$ for all P. We define the divisor 0 to be the divisor for which $n_P = 0$ for all P. For any divisor D, define the vector spaces
$$L(D) = \{f \in K(X) : (f) + D \geq 0\} \subset K(X)$$
and
$$I(D) = \{\eta \in M(X) : (\eta) \geq D\} \subset M(X).$$
We define $\ell(D)$ and $\iota(D)$ to be the dimensions of $L(D)$ and $I(D)$, respectively.

Black Box 2.3.2 (Riemann-Roch [95, Theorem 5.4.1]). *If X is a genus g Riemann surface, $D \in \text{Div}(X)$,*
$$\ell(D) - \iota(D) = \deg(D) + 1 - g.$$

Exercise 2.10. *Use Black Box 2.3.2 to prove that for any holomorphic 1-form ω on a closed Riemann surface X of genus g with zeros of order k_i, $i = 1, \ldots, n$,*
$$\sum_{i=1}^{n} k_i = 2g - 2.$$

2.3. Holomorphic 1-forms

Connection to geometric structures. We now give some basic results from the holomorphic definition, which will prepare us for showing that the three definitions we have outlined are in fact equivalent. We will first show how integrating ω gives natural coordinates whose transition maps are translations, yielding our geometric structure definition. For any point $p \notin \Sigma$, let U_α be a simply connected domain containing p such that $U_\alpha \cap \Sigma = \emptyset$. Let $w_\alpha : U_\alpha \to \mathbb{C}$ be a holomorphic chart. We can assume $w_\alpha(p) = 0$. The corresponding holomorphic function ϕ_{w_α} is nonzero at $w_\alpha(p)$, and by shrinking U_α if necessary, we can assume ϕ_α is nonzero in $w_\alpha(U_\alpha)$. Set

$$z_\alpha(w_\alpha(q)) = \int_0^{w_\alpha(q)} \phi_{w_\alpha}(\tau) d\tau,$$

where the integration is over a path from 0 to $w_\alpha(q)$ in U_α. Since U_α is simply connected, Cauchy's theorem [147, §1.6] implies that the integral is independent of the path, so $z_\alpha(w_\alpha(q))$ is well-defined. Then

$$\frac{dz_\alpha}{dw_\alpha}(w_\alpha(q)) = \phi_{w_\alpha}(w_\alpha(q)) \neq 0,$$

and so z_α defines a new coordinate chart with

$$\phi_{z_\alpha} \equiv 1.$$

In an overlapping chart U_β with coordinate function z_β we have

$$\frac{dz_\beta}{dz_\alpha} \equiv 1$$

so there is a constant c such that

$$z_\beta = z_\alpha + c.$$

These are called the *natural coordinates* associated to ω and in these coordinates the metric is

$$|\omega| = |dz|,$$

the Euclidean metric.

Zeros. Now let $p \in \Sigma$ be a zero of order k of ω. Thus for any coordinate function z with p corresponding to $z = 0$, we can write the holomorphic function ϕ_z as

$$\phi_z(z) = z^k g(z),$$

where g is holomorphic and $g(0) \neq 0$. Then

$$h(z) = \int_0^z \tau^k g(\tau) d\tau$$

has a zero of order $(k+1)$ at 0, so we can take a $(k+1)$st holomorphic root of h. Set

$$w(z) = (k+1)^{\frac{1}{k+1}} h(z)^{\frac{1}{k+1}};$$

then
$$\frac{dw}{dz} = \frac{1}{k+1}(k+1)^{\frac{1}{k+1}} h^{\frac{-k}{k+1}} h'(z) = w^{-k}\phi_z(z).$$
This says in the coordinates w that
$$\phi_w(w) = w^k.$$
Divide the w plane into $2(k+1)$ sectors of angle $\frac{\pi}{k+1}$, four of which have a ray along either the positive or negative real axis. Letting
$$z(w) = \frac{1}{k+1} w^{k+1}$$
each sector is mapped to a Euclidean half-plane. Since $|dz| = |w^k dw|$, the map $z(w)$ is an isometry with respect to the metrics $|w^k dw|$ and the Euclidean metric $|dz|$ in each sector. In polar coordinates the metric is
$$ds^2 = |w^k dw|^2 = r^{2k}(dr^2 + (rd\theta)^2).$$
For each integer j the rays $\arg(w) = \frac{\pi j}{k+1}$ and $\arg(w) = \frac{\pi(j+1)}{k+1}$ are sent to the real line by the map $w \mapsto z(w)$. We therefore recover the half-plane description in the geometric structures definition.

Flat geometry. We next observe that the holomorphic 1-form determines a locally defined metric $|\omega|$ which in the natural coordinates z away from the zero is defined by $|dz|$ and in a neighborhood of a zero of order k is given by $|z^k dz|$. Given an oriented path γ on X, we define the associated *holonomy vector* z_γ by

$$(2.3.2) \qquad z_\gamma = \int_\gamma \omega.$$

Note that this depends only on the homotopy class of γ. We define the length of a path γ by

$$(2.3.3) \qquad \ell(\gamma) = \int_\gamma |\omega|$$

and the area of a set as

$$(2.3.4) \qquad \mathrm{area}(A) = \frac{i}{2} \int_A \omega \wedge \bar{\omega}.$$

Proposition 2.3.3. *In a neighborhood of a point p which is not a zero of ω, a geodesic is a straight line in the natural coordinates. For a geodesic path passing through a point p which is a zero of ω the angle between an incoming line and outgoing line in the same direction is at least π.*

Proof. The first statement is immediate. To prove the second statement, without loss of generality assume that the incoming geodesic to p from a starting point x is along a horizontal line which is on the boundary of two half-planes. The geodesic cannot leave p within an angle less than π of the horizontal line,

2.3. Holomorphic 1-forms

for then it would leave p by entering one of those two half-planes. But that is not a geodesic. There would be a shorter Euclidean line joining x to a point on the line that is contained in that half-plane.

On the other hand we claim that the path starting at x coming in along a horizontal line to p and then leaving along a straight line at angle at least π with the horizontal is in fact a geodesic. Suppose to the contrary the geodesic starting at x leaves one of the two bounding half-planes, denoted $\mathbb{H}_1$ at a point $x_1 \neq p$ where x_1 is on the horizontal line. The geodesic σ_1 joining x and x_1 is horizontal and passes through p. Now x_1 is on the boundary of $\mathbb{H}_1$ and another half-plane denoted $\mathbb{H}_2$. If the endpoint y of the geodesic is in the interior of $\mathbb{H}_2$, then the line in $\mathbb{H}_2$ from p to y is strictly shorter than σ_1 followed by the line from x_1 to y in $\mathbb{H}_2$. We are done in this case.

If on the other hand y is not in the interior of $\mathbb{H}_2$, then in passing from x_1 to y we must leave $\mathbb{H}_2$ at some x_2 on another horizontal line. The geodesic from x to x_2 would be σ_1 followed by the segment σ_2 from x_1 to x_2 that passes back along the part of σ_1 from x_2 to p. Altogether the subsegment of σ_1 from x to p followed by the segment from p to x_2 in $\mathbb{H}_2$ is strictly shorter than σ_1 followed by σ_2. This means the geodesic in fact does not leave $\mathbb{H}_1$ at x_1 but at p. □

Theorem 2.3.4. *Given any two (not necessarily distinct) points p and q and a homotopy class of arcs joining p and q there is a unique geodesic γ in that homotopy class.*

Proof. The existence follows from the Arzela-Ascoli theorem (see, for example, Rudin [**143**]), and we leave its proof as an exercise. We prove uniqueness. Suppose there are two geodesics γ_1, γ_2. Since they are homotopic they bound a disc D. If there are no zeros in D, then the metric is Euclidean in the disc, and we get a contradiction since geodesics are unique in a Euclidean disc. The proof is by induction on the number of zeros in D. Suppose first there is a unique zero p. Pick a direction θ. Since the cone angle is $(2k+2)\pi$ at p there are $2k+2$ lines in direction θ leaving p. These are called *separatrices*. Since $2k+2 \geq 4$ there must be a pair of adjacent separatrices making angle π which both intersect the same γ_i. But then they bound a Euclidean half-plane and a segment of γ_i is a geodesic in that half-plane joining a pair of points on the boundary. This is impossible. A similar argument holds if there is more than a single zero inside. We take a direction θ at one of the zeros such that the separatrices in that direction do not hit another zero and again we see there is γ_i such that two adjacent separatrices hit γ_i. Those two separatrices and a segment of γ bound a disc. If there is no zero inside, we again have a contradiction. If there is, we apply induction to the pair of geodesics, one of which is the segment of γ_i and the other is the geodesic formed by the pair of separatrices. □

Closed curves. Note that in the above definition if we take $p = q$, we get geodesic representatives for homotopy classes of *closed* curves passing through p.

2.4. Saddle connections and cylinders

In this section, we record several important basic definitions about special trajectories on translation surfaces and associated subsets of the complex plane $\mathbb{C}$.

Definition 2.4.1. A *saddle connection* γ on a translation surface ω is a geodesic in the flat metric determined by ω joining two zeros with no zeros in its interior. See Figure 2.9 for an example of a saddle connection on an L-shaped translation surface.

Figure 2.9. A saddle connection on an L-shaped translation surface.

Cylinders. For our next definition, we introduce some notation associated to Euclidean cylinders in $\mathbb{R}^3 = \mathbb{C} \times \mathbb{R}$. For $c, h > 0$, we define the cylinder of circumference (or *width*) c and height h by

$$\mathcal{C}_{c,h} = \left\{ \left(\frac{c}{2\pi} e^{i\theta}, t \right) : 0 \leq \theta < 2\pi, 0 \leq t \leq h \right\}.$$

Definition 2.4.2. A *cylinder* C in a translation surface ω is an isometric image of $\mathcal{C}_{c,h}$. A cylinder can be in any fixed direction of ω. For fixed $0 < t < h$ the image of the circle in the cylinder is a closed geodesic γ_t of length c with respect to the flat metric defined by ω, which we refer to as a *core curve* of the cylinder. As one varies t the loops γ_t are freely homotopic of the same length and parallel in some direction θ_0. We say an (open) cylinder is *maximal* if it cannot be enlarged. The *modulus* μ_C of the maximal cylinder is defined to be $\frac{h}{c}$, the ratio of the height to the circumference.

Definition 2.4.3. Let γ be a path on ω. The *holonomy vector* of γ, as defined in (2.3.2), is

$$z_\gamma = \int_\gamma \omega \in \mathbb{C}.$$

If we write $z_\gamma = r_\gamma e^{i\theta_\gamma}$, the *length* of γ, $\ell(\gamma)$, is given by $r_\gamma \in \mathbb{R}^+$, and the direction of γ is given by θ_γ. In the case that γ is a core curve of a cylinder or a saddle connection, we use these terms to refer to the cylinder or saddle connection itself.

2.4. Saddle connections and cylinders

Boundaries. In a translation surface, the boundaries of (maximal) cylinders are given by saddle connections in the same direction.

Proposition 2.4.4. *The boundary of a maximal cylinder C in direction θ_0 consists of a union of saddle connections in direction θ_0.*

Proof. Given a point p in the interior of C and q on the boundary of C choose a segment $\sigma \subset C$ joining p and q. If we flow σ in direction θ_0, each point of σ other than q closes up in time c, the circumference of C. Flowing q time t we stay a constant distance from the image of points of σ where one flows time t. This would give a closed loop through q unless the flow line hits a zero. Since q is on the boundary the latter must happen. Since this is true for all $q \in \partial C$, the boundary must be a union of saddle connections in direction θ_0. $\square$

Examples of cylinders. Nice examples of cylinders can be found in our earlier examples. We strongly encourage the reader to complete the following two exercises and to draw accompanying pictures:

Exercise 2.11. *Let ω be the translation surface associated to the regular octagon with opposite sides identified. Show that:*

- *The vertical lines joining the top horizontal side to the bottom side close up and determine a cylinder, and the boundary is a pair of vertical saddle connections joining a pair of vertices.*
- *Show that there is another vertical cylinder by taking a vertical line leaving one of the sides making angle $3\pi/4$ with the horizontal, and showing that after a pair of side identifications it closes up and determines a cylinder as well.*
- *Conclude that the surface decomposes into a union of two open maximal cylinders in the vertical direction and vertical saddle connections.*

Exercise 2.12. *Show that every L-shaped table decomposes in the vertical and horizontal directions into the union of two maximal cylinders and parallel saddle connections.*

Geodesic representatives. We observe that in each homotopy class, there is a unique geodesic representative.

Proposition 2.4.5. *For every free homotopy class of a simple closed curve on the underlying surface, there is a geodesic representative on ω. It is unique and is a union of saddle connections except in the case when it is a closed loop contained in the closure of a cylinder.*

Proof. The existence of a length-minimizing path in the homotopy class again comes from taking limits. If it does not pass through a zero, then there must be parallel curves of the same length, sweeping out a cylinder. If a geodesic is not in the closure of a cylinder, we argue it is unique. Otherwise if there were two such geodesics, they would be disjoint by Theorem 2.3.4. Therefore they would bound an annulus. Just as in the proof of Theorem 2.3.4 there could not be interior zeros in the annulus. Since the Euler characteristic of an annulus is zero and both the Gaussian curvature and the geodesic curvature are zero, the Gauss-Bonnet theorem (Black Box 2.2.1) says that the turning angle at each zero on the boundary between incoming and outgoing saddle connections is zero so the angles they make with each other are exactly π, which says that the boundary loops are on the boundary of a Euclidean cylinder, contrary to assumption. □

2.4.1. Sets of holonomy vectors.

Definition 2.4.6. We define $SC(\omega)$ as the set of all saddle connections equipped with an orientation on ω and let

$$(2.4.1) \qquad \Lambda_\omega = \{z_\gamma : \gamma \in SC(\omega)\} \subset \mathbb{C}$$

be the associated set of holonomy vectors. Similarly, define $Cyl(\omega)$ to be the set of (homotopy classes of) core curves of cylinders on ω and

$$(2.4.2) \qquad \Lambda_\omega^{cyl} = \{z_\gamma : \gamma \in Cyl(\omega)\} \subset \mathbb{C}$$

to be the associated set of holonomy vectors.

Lemma 2.4.7. *$SC(\omega)$ is a countable set, and Λ_ω is a countable discrete subset of $\mathbb{C}^*$.*

Proof. Let X denote the Riemann surface underlying ω, and let $\Sigma \subset X$ denote the (finite) set of zeros of ω. Let γ_n be a sequence of saddle connections with holonomy vectors

$$z_n = \int_{\gamma_n} \omega.$$

We argue by contradiction. Suppose there is a $z_\infty \in \mathbb{C}$ so that

$$\lim_{n \to \infty} z_n = z_\infty.$$

By passing to a subsequence if needed, we can assume that the γ_n all start at the same zero $p \in \Sigma$ and end at the same zero $q \in \Sigma$. Since the total angle at p is finite we can assume the angles $\theta_{n,m}$ between γ_n and γ_m at p converge to 0 as $n, m \to \infty$. Consequently, we see that the directions of the γ_n converge to a limiting direction θ_∞ and the geodesic γ_∞ of length $|\gamma_\infty| = |z_\infty|$ in that direction starting at p has holonomy z_∞. We note that γ_∞ might consist of saddle connections followed by a separatrix all of which are in the same direction and making an angle π at any interior zero of γ_∞. Furthermore the γ_n converge uniformly to γ_∞. This will lead to a contradiction. Note that the convergence of γ_n

2.5. Equivalences

to γ_∞ implies that q must be the other endpoint of γ_∞. But this is a contradiction since for any saddle connection γ_0 (or a collection of segments as above) a straight line leaving p in an angle close to that of γ_0 diverges from γ_0 and does not hit the same zero q in distance close to $|\gamma_0|$. Applying this to γ_∞ gives a contradiction to the convergence of z_n to z_∞. Since $z_\infty \in \mathbb{C}$ was arbitrary, we have that Λ_ω is discrete. □

Homologous saddle connections. We note that the map $\gamma \mapsto z_\gamma$ is not necessarily a bijection between SC(ω) and Λ_ω, as multiple saddle connections can have the same holonomy vector. This happens in particular if two saddle connections are *homologous*. An example of homologous saddle connections is the slits in Figure 2.5. The union of the two slits separate the pair of tori.

Cylinder holonomies. The set of cylinder holonomies is also discrete. We leave the proof as an exercise:

Exercise 2.13. *Use Lemma 2.4.7 and Proposition 2.4.4 to prove that $\Lambda^{cyl}(\omega)$ is a discrete subset of $\mathbb{C}^*$.*

Examples. While the square torus $(\mathbb{C}/\mathbb{Z}[i], dz)$ does not have any saddle connections (since the 1-form dz does not have any zeros), we can consider the set Λ of holonomy vectors of simple closed geodesics passing through 0.

Exercise 2.14. *Show that for Λ as defined above*
$$\Lambda = \mathbb{Z}_{\text{prim}}[i] = \{m + ni : m, n \in \mathbb{Z}, \gcd(m, n) = 1\}.$$

2.5. Equivalences

In the section, we prove the equivalence of the three definitions of translation surfaces. We record the schematics of this proof in Figure 2.10.

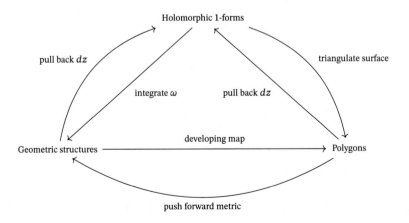

Figure 2.10. A schematic of the equivalences of the definitions.

2.5.1. Polygons to geometric structures.
We first show how to go from a polygon description of a translation surface to a geometric structure description, essentially by using the polygons to give us charts. Away from the vertices the embedding of the polygon in the plane allows for a coordinate chart z coming from the plane. Since the edge identifications are translations, the overlap maps are translations, as desired. By Lemma 2.1.2 the angle around an identified vertex is $2\pi(k+1)$ for a positive integer k. Let the vertex correspond to 0 in the complex plane. Start with a line in the direction of the positive real axis in some polygon. Travel angle π in the counterclockwise direction through polygons glued at the vertex. This sweeps out a half-plane. Continue in a counterclockwise direction another angle π. This allows one to glue a lower half-plane along its negative real axis to the negative real axis of the first plane. We continue gluing $2 + 2k$ half-planes until we return to the original line. This gives us the picture of the neighborhood of the vertex in the geometric structure definition.

2.5.2. Geometric structures to holomorphic forms.
Next, we show how to get a Riemann surface and a holomorphic 1-form on our surface from an atlas of charts. The idea here is to pull back dz from $\mathbb{C}$ using charts. Again, there are issues at the singular points. At each singular point, there are $2k+2$ half-planes indexed by $0 \leq j \leq 2k+1$ for some positive integer k. Map each plane to a sector in the upper half-plane with angle $\frac{\pi}{k+1}$ with one boundary on the positive real axis by a map

$$w(z) = ((k+1)z)^{\frac{1}{k+1}}$$

for an appropriate $(k+1)$st root. Then follow by rotation that takes this sector to a sector with angle $\frac{\pi}{k+1}$ bounded by lines through the origin at angles $\frac{\pi j}{k+1}$ and $\frac{\pi(j+1)}{k+1}$. The coordinates w in each sector fit together to give a holomorphic coordinate w in a neighborhood of 0. The inverse map is

$$z(w) = \frac{1}{k+1} w^{k+1}$$

which maps each sector to a half-plane and if we define the 1-form in a neighborhood of 0 by $w^k dw$, then under the change of coordinates we get that the 1-form is dz in the half-plane, as desired.

2.5.3. Holomorphic forms to polygons.
Finally, we show how to go from a holomorphic 1-form to a polygonal description of the surface, by building a triangulation of the surface whose edges are saddle connections.

Proposition 2.5.1. *Any translation surface ω has a triangulation whose edges are saddle connections.*

2.5. Equivalences

Proof. We construct such a triangulation inductively. First, find an initial saddle connection (which exists since there are geodesics between any two zeros on the surface). Now suppose inductively that we have already constructed a collection of saddle connections $\mathcal{E}$, disjoint except at the zeros. Let Ω be a complementary domain of $\mathcal{E}$ which is not a triangle. If Ω is simply connected, then it is a polygon in the plane which is not a triangle and we can find a segment in Ω joining two vertices. We add that saddle connection to $\mathcal{E}$.

Now suppose Ω is connected but not simply connected. Choose a homotopy class γ of a segment in Ω that is not homotopic to a segment on the boundary and joins a pair of vertices. Replace γ by the geodesic in its homotopy class and let γ_1 be its initial segment with one endpoint denoted p. Since Ω is connected, γ_1 does not begin by entering a triangle with sides in $\mathcal{E}$. If γ_1 is contained in Ω, add it and we are done. If not, then an initial subsegment γ_1' of γ_1 leaves Ω for the first time with endpoint on an edge $e_1 \in \mathcal{E}$. Denote the intersection point by q. Let p' and p'' be the singularities which are the endpoints of e_1 and let β' and β'' be the segments of e_1 from q to p' and p'', respectively. Consider the geodesics κ' and κ'' in the homotopy class of $\gamma_1' * \beta'$ and $\gamma_1' * \beta''$ (where $*$ denotes concatenation) joining p_1 to p' and p'', respectively. Now the edges of κ', κ'', and $\beta' \cup \beta''$ bound a polygon. If it is a triangle, either κ or κ' is not in $\mathcal{E}$ and we can add it. If it is not a triangle, it contains an edge not in $\mathcal{E}$ and we can add it. $\square$

Conclusions. The fact that any surface ω defined by a holomorphic 1-form can be triangulated means that it can be represented by polygons. In a later chapter we will investigate a canonical triangulation, called the *Delaunay triangulation*. We have thus shown how to move between the three different definitions of translation surfaces.

2.5.4. Equivalences of translation surfaces.
We now, in preparation for defining moduli spaces of (equivalence classes of) translation surfaces in Chapter 3, define when two translation surfaces are equivalent, using both the complex analytic and polygonal definitions. We leave how to formulate the equivalence in the geometric structures definition as a (nontrivial) exercise.

Biholomorphisms. In the complex analytic version we say ω_1 is equivalent to ω_2 if there is a biholomorphic map $f : X_1 \to X_2$ such that $f^*\omega_2 = \omega_1$. Here X_i is the underlying Riemann surface for ω_i.

Cut and paste. In the polygon version, we start with a collection of polygons $P = \{P_1, \ldots, P_n\}$, $P_i \subset \mathbb{C}$, with edges to be identified by translation. If an edge of P_i is identified with an edge e of a distinct P_j, we may glue P_i to P_j along e to form a polygon P'. We can also take a polygon P and cut along some line joining vertices to form a pair of polygons with the cut line identified; see

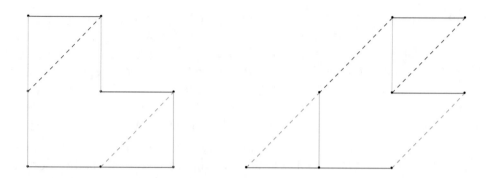

Figure 2.11. The two translation surfaces are equivalent, via cutting and pasting. They give a genus 2 surface with a single zero.

Figure 2.11. In general it might be quite hard to decide when two polygons determine the same translation surface under this equivalence of cutting and pasting.

Equivalence of equivalences. If two translation surfaces are equivalent under cutting and pasting, they are equivalent under holomorphic equivalence, since the cutting and pasting operations do not change the underlying holomorphic structure. Suppose conversely ω_1 is equivalent to ω_2 under the holomorphic equivalence. Triangulate each surface and use the biholomorphic map to map the triangulation of the surface X_2 underlying ω_2 to a triangulation of the surface X_1 underlying ω_1. This gives a pair of triangulations of the same surface X_1.

Flips. Starting with one triangulation and a pair of triangles sharing a side, the union is a quadrilateral. Replacing the shared side with the other diagonal of the quadrilateral, we have a new triangulation. This is called a *flip* move. Given a fixed set of singularities on the surface one can form a graph. The vertices of the graph are triangulations of the surface. The vertices of each triangle are at the finite set of points and two vertices of the graph (namely a pair of triangulations) are connected by an edge if the triangulations are related by a flip move. To conclude that ω_1 and ω_2 are equivalent under cutting and pasting, we need the following black box:

Black Box 2.5.2 ([134]). *The flip graph of a closed surface is connected.*

Geometric structures equivalence. We leave it as a (nontrivial) exercise to formulate a notion of equivalence for geometric structures and to show its equivalence to these other notions.

Exercise 2.15 (Hard). *Formulate a notion of equivalence for translation surfaces using the geometric structures definition that is equivalent to the equivalences defined above. See, for example, Quint* [139].

2.6. Quadratic differentials

We conclude this chapter with a discussion of *quadratic differentials*, also known as *half-translation surfaces*. These objects are generalizations of translation surfaces which we will need to refer to from time to time in the sequel and for which much of the machinery we've developed will work but requires some technical details that for the sake of exposition we would prefer to avoid. As with Abelian differentials (holomorphic 1-forms), there are three equivalent definitions. For further details on quadratic differentials and in particular the connection to Teichmüller theory and classical complex analysis, see the books of Gardiner [77] and Strebel [157].

2.6.1. Polygons. As with Abelian differentials, one can define quadratic differentials using polygons and gluing, though now we must allow rotations by angle π in addition to translations: given a finite collection of polygons $P = \{P_1, \ldots, P_n\}$, $P_i \subset \mathbb{C}$, in the plane $\mathbb{C}$ such that the collection S of sides of P are grouped into pairs (s_1, s_2), $s_i \in S$, such that each side s of P belongs to exactly one pair and the two sides (s_1, s_2) in each pair are parallel and of the same length.

Gluing with rotations. Unlike the Abelian case, we are now allowed to identify the sides via either the Euclidean translation or the composition of the Euclidean translation and the rotation by π. We require that if s_1 is glued to s_2, then moving along the glued side one polygon is to the left of the side and the other to the right of the side. This again guarantees the result is a closed orientable surface which we will denote by $S = (X, q)$. Since this construction involves gluing with translations and rotation by π (i.e., multiplication by -1), which are both holomorphic, it yields a Riemann surface structure X, together with a holomorphic quadratic differential coming from pulling back $(dz)^2$ (since $(d(\pm z + c))^2 = (dz)^2$). Again, we distinguish between a collection of polygons P and the rotated copy $e^{i\theta}P$ (if the original surface yields $S = (X, q)$, the rotated surface will yield the pair $S_\theta = (X, e^{i\theta}q)$).

Flat metrics and foliations. Since translations and rotations are Euclidean isometries, the surface S inherits a local Euclidean metric from the polygons, except possibly at identified vertices. Since translations and rotations by π preserve lines of a fixed slope the family of Euclidean lines of a given slope on the surface through any point that is not a singularity induces a singular foliation of the surface S. Note that because of rotations by π, directions along these lines are no longer well-defined, so we no longer have a well-defined flow in a fixed direction. Rather, for each angle $\theta \in [0, \pi)$ there is a foliation of the surface by lines that make angle θ with the real axis. One can still consider the dynamics of such foliations in much the same way we will consider flows on translation surfaces.

Cone angles. As in the case of Abelian differentials, at any vertex we can take a point on a side near the vertex and traverse a small circle around the vertex following it by the side identifications until it returns. The cone angle is the total angle traversed by loop.

Exercise 2.16. *Show that the cone angle for the metric associated to a quadratic differential is an integer multiple of π.*

2.6.2. Geometric structures. A *half-translation structure* is given by the following data:

Regular points. On $M \setminus \Sigma$ we have an atlas of coordinate charts U_z and coordinate functions

$$\phi_z : U_z \to \mathbb{C}$$

such that for $U_z \cap U_w \neq \emptyset$, the transition function

$$\phi_z \circ \phi_w^{-1} : \phi_w(U_z \cap U_w) \to \phi_z(U_z \cap U_w)$$

is a *half-translation*, that is of the form

$$\phi_z \circ \phi_w^{-1}(\zeta) = \pm \zeta + c$$

for a constant $c \in \mathbb{C}$. The terminology *half-translation* is explained by the fact that the squares of these maps are translations. As in the Abelian setting, we can build the local Euclidean metric $|dz| = |d(-z)|$ on $M \setminus \Sigma$.

Singular points. For $p \in \Sigma$, the structure around p is as follows. For a positive integer $\ell \geq 3$ take ℓ Euclidean (upper) half-planes $\mathbb{H}_j$, each bounded by the real axis. In the case that $\ell = 2k + 2$ is even the situation is exactly as in the Abelian case. Recall that this means to glue all the half-planes at 0 and to glue their boundaries as follows: For j odd the half-planes have positive y-coordinate and for j even, negative y-coordinate. For j odd glue the negative real axis of $\mathbb{H}_j$ to the negative real axis of $\mathbb{H}_{j+1}$ and for j even glue the positive real axis of $\mathbb{H}_j$ to the positive real axis of $\mathbb{H}_{j+1}$ with the indices of the half-planes viewed mod ℓ, so $\mathbb{H}_{2k+3} = \mathbb{H}_1$.

Odd ℓ. The new possibility is that $\ell = 2k + 1$ is odd and there are ℓ half-planes. In that case the gluings are the same as in the Abelian case for $j \leq 2k$, but now the negative real axis of $\mathbb{H}_\ell$ is glued to the positive real axis of $\mathbb{H}_1$ by a π rotation. The angle traced out by a loop around 0 is $(2k+1)\pi$, with $(2k-1)\pi$ being the excess angle. The metric in this neighborhood of $p \in \Sigma$ is given once again by choosing polar coordinates (r, θ) and defining a metric

$$ds^2 = r^{\frac{\ell}{2}-1}(dr^2 + (rd\theta)^2).$$

2.6. Quadratic differentials

2.6.3. Complex analysis. A *holomorphic quadratic differential* q on a Riemann surface X is a tensor of the form $f(z)(dz)^2$ in local coordinates, where $f(z)$ is holomorphic. That is, q assigns to each local coordinate z_α a holomorphic function $\phi_{z_\alpha}(z_\alpha)$ such that in the overlapping charts

$$(2.6.1) \qquad \phi_{z_\beta}(z_\beta)\left(\frac{dz_\beta}{dz_\alpha}\right)^2 = \phi_{z_\alpha}(z_\alpha).$$

Any Abelian differential ω on X yields a quadratic differential $q = \omega^2$ on X. This corresponds in the polygon setting to gluings which do not involve any rotations, and in the geometric structures case to transition maps which do not have rotations.

Sections. In the language of algebraic geometry, a quadratic differential is a section of the *square* $K^2(X)$ of the canonical (i.e., holomorphic cotangent) bundle $K(X)$ of X. The set of holomorphic quadratic differentials on X is a $\mathbb{C}$-vector space $Q(X)$, and we have the following:

Black Box 2.6.1. $Q(X)$ *is a complex vector space of dimension* $3g - 3$; *see, for example,* [95, Corollary 5.4.2].

Zeros and cone angles. As in the Abelian setting, there is a connection between the orders of zeros of the differentials, cone angles, and genus, which we leave as an exercise.

Exercise 2.17. *Show that if q has a zero of order k at p_0 (i.e., in a neighborhood of p_0 $q = z^k(dz)^2$), then the total angle of the flat metric at p_0 is $(k + 2)\pi$. Show that if X has genus g, the sum of the orders of zeros of q is $4g - 4$.*

2.6.4. Double covers. Associated to any quadratic differential (X, q) is a canonical double cover $(\tilde{X}, \tilde{q})$. To describe a cover one needs to identify the subgroup of closed curves in the fundamental group that lift to closed curves. In this case remove the singularities and declare that a loop in the complement lifts to a closed loop, if for any direction, say horizontal, in traversing the loop a vector pointing in the positive horizontal direction returns to itself and not its negative. Loops around singularities have this property if and only if the total angle is an even multiple of π, or in other words if the singularity corresponds to a zero of even order. The lifted differential $\tilde{q} = \omega_q^2$ is the square of an Abelian differential ω_q. The cover is branched at the (possibly empty) finite set of zeros of odd order ℓ. At such points with cone angle $\pi(\ell + 2)$ the cone angle on the double cover is $2\pi(\ell + 2) = 2\pi(\ell + 1 + 1)$ and corresponds to a zero of order $\ell + 1$ of the Abelian differential ω_q. There is an involution τ of the double cover satisfying $\tau^*\omega_q = -\omega_q$ which exchanges the two sheets of the cover.

Tiling by polygons. The covering construction is perhaps easiest to see in the polygon definition, where the construction is very explicit: from the collection of polygons $P = \{P_i\}$, form the collection $\tilde{P} = \{\pm P_i\}$, and now each side has a companion side that can be glued only using translations. Note that there will be two possible choices of vertical direction, corresponding to the square roots $\pm\omega_q$ of the lifted differential $\tilde{q}$.

Area. Note that since

$$(dz \wedge d\bar{z}) = (d(-z) \wedge d(\overline{-z})),$$

the area form

$$\left(q^{1/2} \wedge \bar{q}^{1/2}\right)$$

is well-defined even if a global square root of q is not well-defined.

Exercise 2.18. *Show that*

$$\mathrm{Area}(\tilde{X}, \tilde{q}) = 2\,\mathrm{Area}(X, q).$$

Meromorphic differentials. When considering quadratic differentials, it is often useful to consider *meromorphic* differentials. In particular, quadratic differentials with at most simple poles have finite area, and the Abelian differential on the canonical double cover is holomorphic. A simple pole corresponds to a cone point with angle π, that is, a deficiency of angle, which can be viewed as local *positive* curvature.

Exercise 2.19. *Consider the regular tetrahedron with its natural flat metric. Show that this comes from a quadratic differential on $\mathbb{C}P^1$ with four simple poles corresponding to the vertices, and show that the double cover yields a torus whose underlying parallelogram is a rhombus tiled by two equilateral triangles. Hint: Use the classical net for the tetrahedron.*

2.6.5. Higher-order differentials. We can also consider differentials of orders integers $j \geq 2$, which correspond to polygons with gluings by angle $2\pi/j$ and translations; geometric structures with rotations by $2\pi/j$ and translations; and tensors of the form $f(z)(dz)^j$ (or sections of the jth power $K^j(X)$ of the canonical bundle of the underlying Riemann surface X) in the polygon, geometric structure, and complex analytic picture, respectively. Associated to a j-differential is a canonical degree j branched cover (again, branched over zeros and poles) where the differential lifts to the jth power of an Abelian differential. A zero of order k for the j-differential corresponds to an angle of $2\pi\left(1 + \frac{k}{j}\right)$, and on the j-cover, yields an angle of $2\pi(k + j)$, so a zero of order $k + j - 1$ for the Abelian differential.

Exercise 2.20. *If X has genus g, show that the sum of the orders of the zeros (and poles, if we consider meromorphic differentials) for a j-differential on X is $j(2g - 2)$.*

2.6. Quadratic differentials

Platonic solids. A rich source of examples of j-differentials on the Riemann sphere $\mathbb{C}P^1$ is polyhedra whose angles are rational multiples of π. In particular, the Platonic solids give examples of j-differentials for several different k, as you can show via the exercise below.

Exercise 2.21 ([6]). *Show that the flat metrics on Platonic solids correspond to j-differentials with simple poles at their vertices ($j = 2$ for the tetrahedron, $j = 3$ for the octahedron, $j = 4$ for the cube, $j = 6$ for the icosahedron, and $j = 10$ for the dodecahedron). Compute the genus of the canonical j-covers.*

Chapter 3

Moduli Spaces of Translation Surfaces

In this chapter, we introduce the moduli space of translation surfaces. We start by giving a brief introduction to Teichmüller and moduli spaces of Riemann surfaces in §3.1. We then introduce the moduli space of translation surfaces and discuss their stratification by the orders of zeros in §3.2. We show how to build coordinates on these strata by computing periods of the differential in §3.3, and we introduce the $GL^+(2,\mathbb{R})$-action on a stratum Ω (and the associated $SL(2,\mathbb{R})$-action on a stratum $\mathcal{H}$ of unit-area surfaces). We show in §3.4 how to build a natural $SL(2,\mathbb{R})$-invariant measure on $\mathcal{H}$ using period coordinates and a cone construction inspired by Haar measures on $SL(n,\mathbb{R})$. We discuss the basic topology of strata and their connected components in §3.5.1, and we conclude this chapter by introducing *Delaunay triangulations* in §3.6 and using them to prove the finiteness of the measure in §3.7.

3.1. Teichmüller and moduli spaces

We begin with a brief introduction to the Teichmüller and moduli spaces of Riemann surfaces. We will not include many details in this introduction, instead referring the interested reader to the books of Farb-Margalit [**64**] and Hubbard [**89**]. Fix a closed oriented topological surface S of genus $g \geq 1$.

Teichmüller space. The Teichmüller space T_g of genus g is the set of equivalence classes of *marked genus g Riemann surfaces*. A *marked Riemann surface* is a pair (X, f) where $f : S \to X$ is a homeomorphism (called the *marking*) and X is a Riemann surface. We say $(X, f_1) \sim (Y, f_2)$ if there is a biholomorphic map $h : X \to Y$ such that $f_2^{-1} \circ h \circ f_1$ is homotopic to the identity map on S. By a slight abuse of notation, we denote points of Teichmüller space by representatives (X, f).

Mapping class group. The *mapping class group* of S is defined to be

$$\mathrm{Mod}(S) = \mathrm{Diff}^+(S)/\mathrm{Diff}_0^+(S)$$

which is the group of orientation-preserving diffeormorphisms of S ($\mathrm{Diff}^+(S)$) modulo those isotopic to the identity ($\mathrm{Diff}_0^+(S)$)).

Black Box 3.1.1 ([64]). *The group $\mathrm{Mod}(S)$ is a discrete, finitely generated group.*

The action of $\mathrm{Mod}(S)$. The group $\mathrm{Mod}(S)$ acts on Teichmüller space T_g by changing the marking: Given $\phi \in \mathrm{Mod}(S)$ and $(X, f) \in T_g$ let

$$\phi \cdot (X, f) = (X, f \circ \phi^{-1}).$$

Black Box 3.1.2 ([64]). *The mapping class group $\mathrm{Mod}(S)$ acts properly discontinuously on T_g, and the resulting quotient $T_g/\mathrm{Mod}(S)$ is the* Riemann moduli space $\mathcal{M}_g$, *consisting of Riemann surfaces considered up to biholomorphisms. Acting properly discontinuously means for any compact $K \subset T_g$, the set of elements $\gamma \in \mathrm{Mod}(S)$ such that $\gamma(K) \cap K \ne \emptyset$ is finite.*

Black Box 3.1.3 ([64, Theorem 13.2]). *Nielsen [137] and Thurston [159] classified the elements $\phi \in \mathrm{Mod}(S)$ as either* finite order, reducible, *or* pseudo-Anosov. *Reducible means there is a collection Γ of disjoint simple closed curves such that $\phi(\Gamma) = \Gamma$. A pseudo-Anosov element is neither finite order nor reducible and has associated to it a pair of transverse measured foliations (see §5.1.1) where it acts by expansion along one foliation and contraction along the other.*

Dehn twists. For translation surfaces, a particularly important type of reducible element of the mapping class group is *Dehn twists* about simple closed curves α. Let α be a simple closed curve on a surface S and let $U_\alpha = S^1 \times [0, 1]$ be a regular neighborhood of α. We give coordinates for U_α by (θ, r) where $0 \le \theta \le 2\pi$ and $0 \le r \le 1$. The *left Dehn twist* $D_\alpha : U_\alpha \to U_\alpha$ is defined as

$$D_\alpha(\theta, r) = (\theta + 2\pi r, r).$$

The map is the identity on the boundary components $r = 0$ and $r = 1$ and twists a curve crossing U_α to the left. The map is defined to be the identity on the complement of U_α. It is classical (see, for example, [64, 3.1]) that the isotopy class of D_α does not depend on the representative of α nor the neighborhood U_α. Hence it defines an element of the mapping class group $\mathrm{Mod}(S)$.

3.1. Teichmüller and moduli spaces

3.1.1. Metrics and quasiconformal maps. There are several important ways of measuring distances between marked Riemann surfaces, which yield metrics on T_g on which $\mathrm{Mod}(S)$ acts by isometries. The metric that is most closely related to the subject of translation surfaces is the *Teichmüller metric*. In order to define it we first define the notion of a *quasiconformal map*. Suppose $f : X \to Y$ is a homeomorphism of Riemann surfaces and to simplify the discussion, assume that f is smooth except at finitely many points. At each point z where f is differentiable, the derivative $Df(z)$ is an $\mathbb{R}$-linear map from the tangent space $T_z X$ to the tangent space $T_{f(z)}Y$. If the derivative is invertible, it takes concentric circles in the tangent space $T_z X$ of X at z to concentric ellipses in the tangent space $T_{f(z)}Y$ of Y at $f(z)$. Define $K_f(z)$ as the ratio of the major axis to the minor axis of these ellipses and define the *quasiconformal dilatation* of f by

$$K(f) = \sup_{z \in X} K_f(z).$$

Exercise 3.1. *Prove that if f is conformal, then $K(f) = 1$. Use the fact that if f is conformal, the derivative $Df(z)$ is a $\mathbb{C}$-linear map of tangent spaces.*

The Teichmüller metric. The notion of quasiconformality gives us a way to measure quantitatively how different conformal structures are, and it motivates the definition of the *Teichmüller metric*:

Definition 3.1.4. The Teichmüller metric $d_T(\cdot, \cdot)$ is defined as follows:

$$d_T((X, f), (Y, g)) = \inf_h \frac{1}{2} \log K(h),$$

where the infimum is taken over $h : X \to Y$ where $h \circ f \sim g$.

Isometries. We have the following theorem of Royden [141].

Black Box 3.1.5 ([141]). *$\mathrm{Mod}(S)$ is the full group of isometries of T_g equipped with the Teichmüller metric.*

The torus. We start by reviewing the case of the torus, from the introduction and Chapter 2. We fix our smooth torus to be

$$S = S^1 \times S^1 = \{(e^{i\theta_1}, e^{i\theta_2}), 0 \leq \theta_j < 2\pi)\},$$

with the natural smooth structure. To build a (conformal class of) a Riemann surface and an associated marking, we consider quotients of the form $\mathbb{C}/\Lambda$, where Λ is a lattice in $\mathbb{R}^2$.

Markings. To construct a marking, we first choose a lattice and then identify a basis for the lattice. Then we map the standard generators of S to the basis vectors. In defining a lattice up to conformal equivalence (that is, the action of $\mathbb{C}^*$ by multiplication) we can normalize one of the generators to be 1, so the

lattice is generated by $z \to z + 1$ and $z \to z + \tau$ where we choose τ so that $\operatorname{Im} \tau > 0$. We denote this lattice by $\Lambda_\tau = \{m + n\tau : m, n \in \mathbb{Z}\}$. We write
$$\mathbb{T}_\tau^2 = \mathbb{C}/\langle 1, \tau \rangle.$$
We describe the marking for $\tau = i$. The marking $\phi_i : S \to \mathbb{C}/\mathbb{Z}[i]$ is given by
$$\phi_i(e^{i\theta_1}, e^{i\theta_2}) = \left[\left(\frac{1}{2\pi}(\theta_1 + i\theta_2)\right)\right],$$
where as in Chapter 1 we use $[z]$ to denote the coset $z + \mathbb{Z}[i]$.

Exercise 3.2. *Show that ϕ_i is a homeomorphism. Show how to define a similar ϕ_τ to map S to $\mathbb{C}/\Lambda_\tau$ for any $\tau \in \mathbb{H}^2$, and show that for distinct τ the markings are different. Conclude that the Teichmüller space T_1 of tori can be identified with the upper half-plane $\mathbb{H}^2$.*

Exercise 3.3. *If we replace τ with $\tau + 1$, show the marking changes by a composition by a Dehn twist about the vertical curve $(1, e^{i\theta_2})$ in S.*

A geometric marking. To visualize a marking more clearly, one can instead start with the smooth model torus $S \subset \mathbb{R}^3 = \{(x, y, t) : x, y, t \in \mathbb{R}\}$ given by the surface of revolution (about the t-axis) of the circle of radius 1 in the (x, t)-plane centered at $(2, 0, 0)$; that is, the surface
$$S = \{(\cos\theta_1(2 + \cos\theta_2), \sin\theta_1(2 + \cos\theta_2), \sin\theta_2) : 0 \le \theta_j < 2\pi\}.$$
A good exercise is to check that the cross section of S over the plane $y/x = \tan\theta_1$ is indeed the circle of radius 1 centered at $(2\cos\theta_1, 2\sin\theta_1, 0)$. We define the marking $\phi : S \to \mathbb{C}/\mathbb{Z}[i]$ by
$$\phi(\cos\theta_1(2+\cos\theta_2), \sin\theta_1(2+\cos\theta_2), \sin\theta_2) = \left[\left(\frac{1}{2\pi}(\theta_1 + i\theta_2)\right)\right].$$

Exercise 3.4. *Show that ϕ is a homeomorphism, and compute the preimages of the closed curves*
$$\{[x] : 0 \le x \le 1\},$$
$$\{[iy] : 0 \le y \le 1\},$$
and
$$\{[x(1+i)] : 0 \le x \le 1\}$$
in $\mathbb{C}/\mathbb{Z}[i]$. Draw a picture of S and these curves on S. Show how to modify this construction to the torus $\mathbb{C}/\Lambda_\tau$, and draw similar curves on this surface.

Grötzsch's theorem and the Teichmüller metric. An important theorem of Grötzsch (see Ahlfors [2] or Farb-Margalit [64, §11.5]) shows that the optimal quasiconformal map f (the one that minimizes $K(f)$) from one

3.1. Teichmüller and moduli spaces

rectangle to another that sends vertices to vertices is the $\mathbb{R}$-linear map. Precisely, we have

Black Box 3.1.6. *For $a, b > 0$, let*

$$R_{a,b} = \{w = u + iv \in \mathbb{C} : u \in [0, a], v \in [0, b]\}$$

be the rectangle with side lengths a and b. Then the $\mathbb{R}$-linear map $A_{a,b}$ defined by

$$A_{a,b}(x + iy) = ax + iby$$

is the bijection between the square $R_{1,1}$ and $R_{a,b}$ with minimal quasiconformal dilatation.

Exercise 3.5. *Compute the quasiconformal dilatation of the map $A_{a,b}$. Hint: What is the image of a circle under $A_{a,b}$?*

Teichmüller and hyperbolic metrics. Using Exercise 3.5 and the description of T_1 as the upper half-plane $\mathbb{H}^2$, we have the following (challenging) exercise:

Exercise 3.6 (Hard). *Show that the Teichmüller metric on the Teichmüller space $T_1 = \mathbb{H}^2$ is the hyperbolic metric. See, for example, Farb-Margalit [**64**, §11.8.3].*

The mapping class group. In the torus case, the mapping class group can be directly computed.

Exercise 3.7. *Show that the linear action of the matrices*

$$\begin{pmatrix} 1 & 1 \\ 0 & 1 \end{pmatrix} \quad \text{and} \quad \begin{pmatrix} 1 & 0 \\ 1 & 1 \end{pmatrix}$$

yields elements of the mapping class group. Conclude that the mapping class group of the torus contains the group $SL(2, \mathbb{Z})$ of 2 by 2 integer matrices with determinant 1. Show how the curves in Exercise 3.4 transform under these transformations.

Black Box 3.1.7 ([**64**, Theorem 2.5]). *The mapping class group of the torus is the group $SL(2, \mathbb{Z})$.*

Mobius transformations. The group $SL(2, \mathbb{Z})$ acts on $\mathbb{H}^2$ by Mobius transformations,

$$z \mapsto \frac{az + b}{cz + d}.$$

The quotient moduli space $\mathcal{M}_1 = \mathbb{H}^2 / SL(2, \mathbb{Z})$ is what is known classically as the *modular curve*. Another way to see this is that we can always scale our lattice Λ so that the shortest nonzero vector becomes 1, and we choose τ' to be a second shortest vector. By adding the first vector to this second basis vector, that is, shift it horizontally by units of 1, we can choose it so it has real part bounded (in absolute value) by 1/2. That is, we can always choose a basis $(1, \tau)$ with $|\tau| \geq 1$ and $|\text{Re}(\tau)| \leq 1/2$, which precisely describes a fundamental domain for $SL(2, \mathbb{Z})$ acting on $\mathbb{H}^2$ by Mobius transformations.

3.2. Strata of translation surfaces

In this section we define moduli spaces or strata of translation surfaces. Each point in a stratum will be an equivalence class of translation surfaces, where the equivalence was described in §2.5.4. We formalize this as follows.

The space Ω_g. Over each marked Riemann surface (X, f) in Teichmüller space T_g is the g-dimensional vector space $\Omega(X)$ of holomorphic 1-forms on X. Given an element of the mapping class group $\phi \in \text{Mod}(S)$, we can identify the space of holomorphic 1-forms on (X, f) with those on $(X, f \circ \phi^{-1})$. This quotient under this action defines the total space Ω_g of holomorphic 1-forms or (Abelian differentials) as a (orbifold) vector bundle over the moduli space $\mathcal{M}_g$, whose fiber over $X \in \mathcal{M}_g$ is the vector space $\Omega(X)$. As discussed in the previous chapter, each holomorphic 1-form ω on a Riemann surface has zeros of order $\alpha_1, \ldots, \alpha_n$ with $\sum_{i=1}^n \alpha_i = 2g - 2$. We recall from (2.3.4) that

$$\text{Area}(\omega) = \frac{i}{2} \int_X \omega \wedge \overline{\omega}.$$

Definition 3.2.1. Fix an unordered tuple $\alpha = (\alpha_1, \ldots, \alpha_n)$ of nonnegative integers with $\sum_{i=1}^n \alpha_i = 2g - 2$, and consider the set $\Omega(\alpha) \subset \Omega_g$ of all translation surfaces ω with zeros with tuple α. The set $\Omega(\alpha)$ is called a *stratum*. We may equip it with the subspace topology coming from Ω_g. We define the subset

$$\mathcal{H}(\alpha) = \text{Area}^{-1}(1) \subset \Omega(\alpha)$$

of unit-area translation surfaces.

Remark 3.2.2. The identification of holomorphic 1-forms on (X, f) with those on $(X, f \circ \phi^{-1})$ is the same as the holomorphic equivalence relation given in §2.5.4. We remark that this identification is the same as the polygon equivalence described in that section.

The stratum $\mathcal{H}(\emptyset)$. Over each (equivalence class of) torus $\mathbb{T}^2_\tau$ in $\mathcal{M}_1$ there is a 1-complex parameter family of holomorphic differentials $\omega_t, t \in \mathbb{C}$. In the coordinates z on $\mathbb{T}^2_\tau$, the family of 1-forms can be described by

$$\omega_t = t \, dz.$$

The stratum of unit-area tori is denoted $\mathcal{H}(\emptyset)$.

Exercise 3.8. *Show that the area of the translation surface $(\mathbb{T}^2_\tau, \omega_t)$ is $|t|^2 \, \text{Im}(\tau)$.*

Parallelograms. Another way of describing the stratum of unit-area tori $\mathcal{H}(\emptyset)$ is the space of unit-area parallelograms up to the $\mathbb{R}$-linear action of $SL(2,\mathbb{Z})$, that is, the space $SL(2,\mathbb{R})/SL(2,\mathbb{Z})$. Given a matrix

$$g = \begin{pmatrix} a & b \\ c & d \end{pmatrix}$$

we associate the lattice $g\mathbb{Z}[i] = \{m(a+ci) + n(b+di) : m, n \in \mathbb{Z}\}$ and form the (unit-area) complex torus $\mathbb{C}/g\mathbb{Z}[i]$, together with the 1-form dz. Note that this fixes a direction, and we distinguish between the lattices which differ by rotation. This approach allows one to generalize to the space of flat n-tori (equivalently, the space of n-dimensional unimodular lattices) as we will discuss at the start of Chapter 6, and this family of generalizations inspires many of our counting results.

Genus 2 examples. We recall some of the genus 2 examples from Chapter 2. As discussed in §2.1.1, the octagon with parallel sides identified and the L-shaped translation surface belong to the stratum $\Omega(2)$, and the 10-gon and the 2-tori glued along a slit belong to the stratum $\Omega(1,1)$.

Higher genus examples. In Exercise 2.4, you showed that identifying opposite sides of regular $4g$- and $4g + 2$-gons yields, respectively, surfaces in $\Omega(2g-2)$ and $\Omega(g-1, g-1)$.

3.3. Coordinates on components of strata

We now begin our discussion of general strata. We will show how to define local coordinates on strata, and we will use these to develop the notion of a measure on strata. We fix a connected component Ω of a stratum $\Omega(\alpha)$.

3.3.1. Period coordinates. Period coordinates are local coordinates, defined in a neighborhood of a fixed $\omega_0 \in \Omega$. A concrete way of describing these is to first triangulate the surface ω_0 in such a way that the edges of the triangulation are flat geodesics and the vertices of the triangulation are zeros of ω_0. For $\omega \in \Omega$ close to ω_0 use the same combinatorial triangulation but vary the lengths and directions of the edges. To make this precise let Σ denote the zeros of ω_0, and let $s = |\Sigma|$. Each edge of the triangulation gives an element of $H_1(S, \Sigma, \mathbb{Z})$, the first homology of the surface relative to Σ, which has dimension $N = 2g + s - 1$. Choose a collection of edges that gives a basis $\gamma_1, \ldots, \gamma_N$ for this homology group. Then the collection of N complex numbers given by

$$\left(\int_{\gamma_1} \omega, \ldots, \int_{\gamma_N} \omega \right) \in \mathbb{C}^N$$

gives local coordinates, known as *period coordinates*. Thus N is the complex dimension of the component Ω. That is, the local coordinates are given by viewing ω as a linear functional on relative homology, that is, an element of relative cohomology $H^1(S, \Sigma, \mathbb{C})$. We emphasize again that these are *local* coordinates. For example, two edges of a triangle which are linearly independent in relative homology might collapse to a single edge and then the homology basis has to change in order to define new local coordinates.

Perspectives on period coordinates. Period coordinates and their geometry have been studied from a wide range of perspectives and have been particularly important in helping to understand the neighborhoods of the boundary of the spaces Ω and $\mathcal{H}$. A (very) incomplete list of references includes the work of Douady-Hubbard [48], Hubbard-Masur [88], Masur [118] and Veech [162], and Masur-Smillie [121], and more recently, the work of Bainbridge-Chen-Gendron-Grushevksy-Möller [24].

3.4. Measures

We now show how to use period coordinates to define a measure ν on each component Ω of $\Omega(\alpha)$, a construction introduced in work of Masur [118] and Veech [162], and Masur-Smillie [121]. Given $\omega_0 \in \Omega$, fix a basis $B = \{\beta_1, \ldots, \beta_N\}$ for the relative homology group $H_1(S, \Sigma, \mathbb{Z})$. As we have seen

$$\left(\int_{\beta_1} \omega, \ldots, \int_{\beta_N} \omega \right)$$

gives local coordinates for ω in a neighborhood U of ω_0. This defines a local coordinate map $\Psi : U \to \mathbb{C}^N$ which assigns to each $\omega \in U$ these coordinates. On $\mathbb{C}^N$ we take Lebesgue measure λ normalized so that the unit cube $[0,1]^{2N}$ in $\mathbb{C}^N = \mathbb{R}^{2N}$ has measure 1. We transport λ to a measure ν on U by defining, for any Borel $A \subset U$,

$$\nu(A) = \lambda(\Psi(A)).$$

Our first crucial observation is that this measure is in fact well-defined:

Lemma 3.4.1. *The measure ν is independent of the choice of the basis B of $H_1(S, \Sigma, \mathbb{Z})$.*

Proof. Let $\hat{B} = \{\hat{\beta}_1, \ldots, \hat{\beta}_N\}$ be another basis. The change of basis from B to $\hat{B}$ is given by a real linear map

$$T : H_1(S, \Sigma, \mathbb{Z}) \to H_1(S, \Sigma, \mathbb{Z})$$

with integer coefficients $(c_{i,j})$. We can consider T as a real linear map from $\mathbb{R}^N$ to itself and since it is an invertible map with integer coefficients whose inverse also has integer coefficients, it has determinant ± 1. The new local coordinates defined by $\hat{B}$ are

$$\left(\int_{\hat{\beta}_1} \omega, \ldots, \int_{\hat{\beta}_N} \omega \right) = \left(\sum_{j=1}^N c_{1,j} \int_{\beta_j} \omega, \ldots, \sum_{j=1}^N c_{N,j} \int_{\beta_j} \omega \right).$$

3.4. Measures

Writing each $\int_{\beta_j} \omega$ as a vector $(x_j, y_j) \in \mathbb{R}^2$ we see the right side is the vector

$$(T(x_1, \ldots x_N), T(y_1, \ldots, y_N)) \in \mathbb{R}^N \times \mathbb{R}^N = \mathbb{R}^{2N}.$$

Since T has determinant ± 1, Lebesgue measure defined by the basis $\hat{B}$ coincides with Lebesgue measure defined by B. $\square$

Building a finite measure. The measure ν is infinite. To see this, note that we can scale our differentials by any positive real numbers $t \in \mathbb{R}^+$. To build a measure which is locally finite we will need to change our base space (to account for the $\mathbb{R}^+$-scaling). We recall that we have the area 1 connected component

$$\mathcal{H} = \text{Area}^{-1}(1) = \{\omega \in \Omega : \text{Area}(\omega) = 1\} \subset \Omega$$

where, as we defined above, $\text{Area}(\omega) = \frac{i}{2} \int_X \omega \wedge \bar{\omega}$. We define the measure $\mu_{\mathcal{H}}$ (called MSV measure) on $\mathcal{H}$ as follows: for a subset $A \subset \mathcal{H}$, define

(3.4.1) $$\mu(A) = \nu(C(A)),$$

where

(3.4.2) $$C(A) = \{s\omega : 0 < s \le 1, \omega \in A\}.$$

That is, take the ν-measure of the cone $C(A)$ consisting of those ω with area at most 1 whose area 1 rescaling $(\text{Area}(\omega))^{-1/2} \omega$ belongs to the set A. In §3.6 below we will sketch a proof, following work of Masur-Smillie [121] that this measures is indeed finite; that is,

$$\mu(\mathcal{H}) < \infty.$$

Remark 3.4.2. The set of translation surfaces arising as unfoldings of rational billiards has MSV measure 0 in any stratum. This follows from the fact that the unfolding leads to saddle connections representing different homology classes having the same length. That condition is a positive codimension, and therefore it is a 0-measure set.

Cone constructions. The notion of defining the measure in terms of the measure of a cone in a bigger space can be used to define Haar measure on $SL(n, \mathbb{R})$, the group of $n \times n$ real matrices with determinant 1. Let $M_n(\mathbb{R})$ be the set of $n \times n$ real matrices, which is bijective to $\mathbb{R}^{n^2}$. Put Lebesgue measure λ on $\mathbb{R}^{n^2} = M_n(\mathbb{R})$. Now for any $A \subset SL(n, \mathbb{R})$ define Haar measure by

$$\mu(A) = \lambda\{ta : 0 \le t \le 1, a \in A\}.$$

Exercise 3.9. Show that μ is invariant under the left and right actions of $SL(n, \mathbb{R})$ on itself. Hint: Use the fact that $SL(n, \mathbb{R})$ acting linearly on $\mathbb{R}^n$ preserves Lebesgue measure.

3.4.1. The action. As we have seen in (1.2.1), the group $GL^+(2,\mathbb{R})$ of 2×2 real matrices with positive determinant acts on $\Omega(\emptyset)$ via the $\mathbb{R}$-linear action on $\mathbb{C}$. We now describe this action on every $\Omega(\alpha)$. Our description will be in terms of polygons and then in terms of charts.

Polygons. In the polygon version, where ω is represented by a disjoint collection of polygons $(P_1, \ldots, P_k)$, $P_i \subset \mathbb{C}$, with parallel sides identified by translation, $A\omega$ is given by the collection $(AP_1, \ldots, AP_k)$. The edges of the P_i which were parallel of the same length stay parallel of the same length under linear actions, so the translation surface $A\omega$ is well-defined. Figure 3.1 illustrates an example in the torus case of an action by a unipotent matrix, resulting in an (area-preserving) shear.

Charts. We explain next in terms of the geometric version where the translation surface ω is defined by a collection of charts $\{(U_\beta, z_\beta)\}$ either to $\mathbb{C}$ or to the collection of glued half-planes and where the transition maps are translations; that is, if $z_\beta(U_\beta) \cap z_\alpha(U_\alpha) \neq \emptyset$, the transition map $z_\alpha \circ z_\beta^{-1}$ is a translation on its domain of definition. That is, for $w \in z_\beta(U_\beta) \cap z_\alpha(U_\alpha)$,

$$z_\alpha \circ z_\beta^{-1}(w) = w + c_{\alpha,\beta}.$$

The action is given by post-composition with these charts; that is, given $A \in GL^+(2,\mathbb{R})$, the translation surface $A\omega$ is defined by the charts $\{(U_\beta, A \circ z_\beta)\}$. It is important to note that in the case of the cone points, the post-composition is the linear map in each half-plane. The new transition maps are of the form

$$\begin{aligned}(A \circ z_\alpha) \circ (A \circ z_\beta)^{-1}(Aw) &= A \circ z_\alpha \circ z_\beta^{-1} \circ A^{-1} Aw \\ &= A(z_\alpha \circ z_\beta^{-1})(w) \\ &= A(w + c_{\alpha,\beta}) \\ &= Aw + Ac_{\alpha,\beta}.\end{aligned}$$

Lemma 3.4.3. *The subgroup $SL(2,\mathbb{R})$ of determinant 1 matrices preserves the area 1 locus $\mathcal{H}$, and the induced action on $\mathcal{H}$ preserves the measure μ.*

Proof. To see that $SL(2,\mathbb{R})$ preserves $\mathcal{H}$, we use the polygon presentation of a translation surface. Represent $\omega \in \mathcal{H}$ as a union of polygons $P = \{P_1, \ldots, P_k\}$

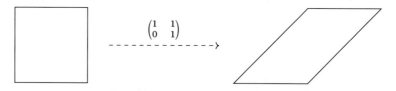

Figure 3.1. A horizontal shear of a torus with a unipotent matrix.

3.4. Measures

glued along edges. The condition $\omega \in \mathcal{H}$ implies the total area $\text{Area}(\omega) = \sum_{i=1}^{k} \text{Area}(P_i) = 1$. If $g \in SL(2, \mathbb{R})$, then the collection of polygons gP determines the surface $g\omega$; therefore, $\text{Area}(g\omega) = \sum_{i=1}^{k} \text{Area}(gP_i) = 1$; therefore, $g\omega \in \mathcal{H}$. To show that the measure μ is preserved, note that the measure ν on Ω is preserved by $SL(2, \mathbb{R})$, since the measure ν is the pullback of Lebesgue measure on $\mathbb{C}^N$, and the action of $SL(2, \mathbb{R})$ in these coordinates is the diagonal $\mathbb{R}$-linear actions, so it preserves Lebesgue measure on $\mathbb{C}^N$. $\square$

Connection to the Teichmüller metric. The connection with the Teichmüller metric is as follows. We consider the 1-parameter subgroup

$$(3.4.3) \qquad A = \left\{ g_t = \begin{pmatrix} e^{t/2} & 0 \\ 0 & e^{-t/2} \end{pmatrix} : t \in \mathbb{R} \right\}.$$

The matrix g_t acts on a translation surface ω by stretching along horizontal lines by a factor of $e^{t/2}$ and by contracting along vertical lines by the same factor. This is called the *Teichmüller geodesic flow*. The terminology comes from the fact that when projected to Teichmüller space or the quotient moduli space, the result is a geodesic with respect to the Teichmüller metric, called a *Teichmüller geodesic*.

Black Box 3.4.4 (Teichmüller's theorem [2, §6]). *The projection of any A-orbit to Teichmüller space is a geodesic with respect to the Teichmüller metric. In fact, all Teichmüller geodesics are projections of A-orbits on the space of holomorphic quadratic differentials. The map defined by an element of A sending an initial point on the geodesic ray to the terminal point is known as a* Teichmüller map.

Other subgroups. We will also make use of the compact subgroup of rotations

$$K = \left\{ r_\theta = \begin{pmatrix} \cos\theta & \sin\theta \\ -\sin\theta & \cos\theta \end{pmatrix} : \theta \in [0, 2\pi) \right\}.$$

We note that $r_\theta \omega = e^{i\theta} \omega$. Another important subgroup is the unipotent 1-parameter subgroup

$$N = \left\{ u_s = \begin{pmatrix} 1 & s \\ 0 & 1 \end{pmatrix} : s \in \mathbb{R} \right\},$$

known as the *Teichmüller horocycle flow*.

Orbits. A major theme in the study of translation surfaces is the interplay between the properties of an individual translation surface and the properties of its orbit under the $SL(2, \mathbb{R})$-action. We will examine these connections in depth in Chapter 5 and Chapter 6.

3.4.2. Higher-order differentials. Implicit in the statement of Teichmüller's thereom (Black Box 3.4.4) is the fact that $GL^+(2,\mathbb{R})$ acts on the moduli space of holomorphic quadratic differentials. This action is once again induced by the $\mathbb{R}$-linear action on $\mathbb{C}$ and is well-defined since the action of the group commutes with multiplication by -1, that is, rotation by $\pi = 2\pi/2$. However for $j \geq 3$, there is no obvious $GL^+(2,\mathbb{R})$-action arising from the $\mathbb{R}$-linear action, as the $\mathbb{R}$-linear action on $\mathbb{C}$ no longer commutes with rotations by $2\pi/j$.

3.5. Components of strata

The strata $\Omega(\alpha)$ need not be connected. The components were classified by Kontsevich-Zorich [107]. We recall their classification here.

3.5.1. Hyperelliptic strata. Recall that a Riemann surface is *hyperelliptic* if there is a 2-fold branched cover

$$\pi : X \to \mathbb{CP}^1.$$

A hyperelliptic Riemann surface is equipped with a *hyperelliptic involution*, a map σ of order 2 which interchanges the sheets of the cover. Every genus 2 surface is hyperelliptic (see, for example, Farkas-Kra [65]). In each genus there are two connected hyperelliptic components of translation surfaces; that is, every surface in the connected component is hyperelliptic. Points of $\Omega_{\text{hyp}}(2g-2)$ (respectively, $\Omega_{\text{hyp}}(g-1,g-1)$) are holomorphic 1-forms ω on hyperelliptic Riemann surfaces which have a single zero of multiplicity $2g-2$ invariant under σ (respectively, a pair of zeros each of order $g-1$ symmetric to each other with respect to the hyperelliptic involution). Note that if a 1-form ω on a hyperelliptic Riemann surface X has a single zero of order $2g-2$, then this zero is necessarily invariant under the hyperelliptic involution σ, because $\sigma^*\omega = -\omega$. Therefore, such an ω must belong to the component $\Omega_{\text{hyp}}(2g-2)$. However, if ω has two zeros each of degree $g-1$, there are two possibilities. The zeros might be interchanged by the hyperelliptic involution, and they might be invariant under the hyperelliptic involution. In the first case ω belongs to the component $\Omega_{\text{hyp}}(g-1,g-1)$. In the second case it does not. For $g = 2$ there are two connected strata, $\Omega(1,1)$ and $\Omega(2)$.

3.5.2. Components of general strata. We begin by choosing a symplectic basis $\{\alpha_i, \beta_i\}$ for the (absolute) first homology $H_1(S, \mathbb{Z}/2\mathbb{Z})$ of the surface with coefficients in $\mathbb{Z}/2\mathbb{Z}$. This means $i(\alpha_i, \alpha_j) = i(\beta_i, \beta_j) = 0$ for all i, j and $i(\alpha_i, \beta_j) = \delta_{i,j}$. Here $i(\cdot, \cdot)$ is the *algebraic* intersection number in first homology; see, for example, [64, §6.1]. Consider a general translation surface ω having zeros of even degrees, and consider a smooth simple closed curve α which does not contain any zeros of ω. As one traverses α the total change of the angle between the vector tangent to α and the vector tangent to the horizontal line is an integer

3.5. Components of strata

multiple n of 2π. We define the *index* of α by $\mathrm{Ind}(\alpha) = n$. The *parity* of ω (more correctly, the parity of the *spin structure* induced by ω) is given by

$$P(\omega) = \sum_{i=1}^{g}(\mathrm{Ind}(\alpha_i) + 1) \cdot (\mathrm{Ind}(\beta_i) + 1) \pmod{2}.$$

The precise definition of a spin structure is beyond the scope of our book; we refer the interested reader to [107]. This parity will be an invariant that allows us to distinguish connected components of strata.

Genus 4 and higher. We first record the classification of these components for genus $g \geq 4$, where the list is simpler than in low genus, which has several exceptional cases.

Theorem 3.5.1 ([107]). *Assume $g \geq 4$. For g odd, the strata $\Omega(2g-2)$ and $\Omega(g-1, g-1)$ each have three connected components: the hyperelliptic component, a component consisting of those ω with even spin structure, and a component with odd spin structure. For g even the same is true for $\Omega(2g-2)$. The stratum $\Omega(g-1, g-1)$ has the hyperelliptic component and one other. Other strata with all even zeros have two components, those with even and those with odd spin structure. Strata with an odd order zero are connected.*

Genus 2 and 3. For genus 2 and 3, the classification is as follows:

Theorem 3.5.2 ([107]). *In genus $g = 2$ the strata $\Omega(2)$ and $\Omega(1,1)$ are hyperelliptic and connected. In genus $g = 3$ the strata $\Omega(4)$ and $\Omega(2,2)$ each have a hyperelliptic component and another component of odd spin structures. All other strata are connected.*

More classification of components. There has been a lot of recent progress in the classification of connected components of strata of quadratic (see the work of Boissy [26] and Lanneau [112]) and k-differentials for $k > 2$ (see, for example, the work of Chen-Gendron [34] and the references within).

Noncompactness. We end this section with the following basic, but crucial, proposition. Note that each component Ω of $\Omega(\alpha)$, as defined, are noncompact, since we have an $\mathbb{R}^+$-action by scaling (so the area function, for example, is continuous and unbounded). We now note that connected components of strata of unit-area translation surfaces are noncompact. We first define a function, the *saddle connection systole*, which will be important for us throughout the sequel.

Definition 3.5.3. Let $\mathcal{H}$ be a connected component of a stratum of unit-area translation surfaces $\mathcal{H}(\alpha)$. We define the *saddle connection systole function* $\ell : \mathcal{H} \to \mathbb{R}^+$ by

$$\ell(\omega) = \min_{z \in \Lambda_\omega} |z|.$$

This function measures the length of the shortest saddle connection on ω.

Properties of ℓ. Note first that for any ω, $\ell(\omega) > 0$, by the discreteness (Lemma 2.4.7) of the set of holonomy vectors $\Lambda_\omega \subset \mathbb{C}^*$. Note that we can define the function ℓ on strata of nonnormalized surfaces and that for $s \in \mathbb{R}^+$, $\ell(s\omega) = s\ell(\omega)$.

Lemma 3.5.4. *The function ℓ is continuous on each Ω and hence on $\mathcal{H}$.*

Proof. We can always choose local period coordinates so that a shortest (not necessarily unique) saddle connection is part of a homology basis. Thus ℓ is a continuous function, since it is smooth in these period coordinates. □

Proposition 3.5.5. *Components $\mathcal{H}$ of strata of unit-area translation surfaces are not compact.*

Proof. Let $\omega \in \mathcal{H}$, and let γ be any saddle connection on ω. Suppose

$$z_\gamma = \int_\gamma \omega = re^{i\theta}.$$

Rotate ω by an angle $\psi = \pi/2 - \theta$ so γ is vertical on $e^{i\psi}\omega$. Apply the Teichmüller flow g_t as $t \to \infty$ to $e^{i\psi}\omega$. Then the length of γ on $g_t e^{i\psi}\omega$ shrinks to 0; hence $\ell(g_t r_\theta \omega) \to 0$. Thus ℓ is a continuous function which does not achieve its minimum, and therefore $\mathcal{H}$ is noncompact. □

3.6. Delaunay triangulations

We now show how to construct a canonical triangulation associated to a translation surface ω, following closely the work of Masur-Smillie [121]. A crucial application of this construction will be the proof that the measure of each stratum is finite. Given a translation surface ω with underlying Riemann surface X, let $\Sigma \subset X$ be the set of zeros of ω. We have a distance $d(\cdot, \cdot)$ between any two points $p_1, p_2 \in X$ defined by

$$d(p_1, p_2) = \inf_\gamma \left\{ \left\| \int_\gamma \omega \right\| \right\},$$

where the infimum is over all paths γ joining p_1 and p_2. Following [121], we define the *diameter* of the translation surface ω to be

$$\mathrm{diam}(\omega) = \max_{p \in X} d(p, \Sigma).$$

To quote from [121], the quantity $\mathrm{diam}(\omega)$ is "related to the standard notion of diameter of a metric space but is more useful for our purposes."

3.6. Delaunay triangulations

Tangent spaces and exponential maps. Let $T_p = T_pX \equiv \mathbb{C}$ be the tangent space to X at p and denote by p' the origin in T_p. Let $\iota_p : T_p \to X$ be the exponential map, so $\iota_p(p') = p$. Let $s_p = d(p, \Sigma)$ and let $D(p', s_p) \subset T_p$ be the disc of radius s_p about p'. Recall that Σ is the set of zeros of ω. Let

$$\Sigma_p = \iota_p^{-1}(\Sigma) \cap \partial D(p', s_p)$$

be the set of points in $\partial D(p', s_p)$ that are mapped to zeros of ω under the exponential map ι_p. Since the metric defined by ω is locally Euclidean, the map ι_p is a local isometric immersion of $D(p', s_p) \subset T_p$ into X. For each point $z \in \Sigma_p$, the line segment $[p', z] \subset D(p', s_p)$ from p' to z is mapped by ι_p to a length-minimizing line from p to $\iota_p(z) \in \Sigma$.

The Voronoi decomposition. We now define the *Voronoi decomposition* of the surface ω.

Definition 3.6.1. The Voronoi 2-facets (faces) are the path components of the set of points p for which there is a unique length-minimizing path to Σ. The Voronoi 1-facets (edges) are the points p for which there are exactly 2 length-minimizing paths to Σ and the 0-facets (vertices) are the points that have 3 or more length-minimizing paths to Σ.

The Delaunay decomposition. By taking the dual of the Voronoi decomposition (i.e., faces of the decomposition are now vertices of the dual), we obtain the *Delaunay decomposition* of the surface.

Definition 3.6.2. The *Delaunay decomposition* is dual to the Voronoi decomposition. That is, vertices of the Delaunay decomposition are faces of the Voronoi decomposition, and edges connect two vertices if the corresponding Voronoi cells share an edge.

Facets. For any $p \in X$ we can describe precisely the Delaunay facet containing p. Let $H_p \subset T_p$ be the convex hull of Σ_p. If p is a Voronoi 0-facet, then H_p is 2-(real)-dimensional and the corresponding Delaunay 2-facet is the image under ι_p of the interior of H_p. If p is in a Voronoi 1-facet, then H_p is a segment and the corresponding Delaunay 1-facet is the image under ι_p of the segment minus its endpoints. The Delaunay 1-facet corresponding to p depends only on the Voronoi 1-facet containing p and not on p itself. If p is in a Voronoi 2-cell, then Σ_p has only one point z and the corresponding Delaunay 0-facet is $\iota_p(z)$.

Exercise 3.10. *Find (and draw) the Voronoi and Delaunay decompositions for the translation surface associated to the regular octagon.*

Properties of the Delaunay decomposition. We record crucial properties of the convex hulls H_p in the following proposition:

Proposition 3.6.3. *For any $p \in X$, the map $\iota_p|_{H_p}$ restricted to H_p is injective. Moreover, if p, q are distinct points in X,*

$$\iota_p(\operatorname{int} H_p) \cap \iota_q(\operatorname{int} H_q) = \emptyset$$

unless H_p, H_q are 1-dimensional and $\iota_p(H_p) = \iota_q(H_q)$.

Proof. We will proceed by contradiction for both parts of the proposition.

Injectivity. First suppose ι_p is not injective on H_p. Then there are points $y_p, x_p \in T_p$ with

$$\iota_p(x_p) = \iota(y_p) = x \in X.$$

The line segment $[p', x_p] \in T_p$ is mapped by ι_p to a geodesic path in X joining p to $x = \iota_p(x_p)$. We can isometrically lift the path $\iota_p([p', x_p])$ to T_x, so that x lifts to x', the origin in T_x. Let $p_x \in T_x$ denote the lift to T_x of the endpoint p of this path. The line segment $[p', y_p]$ is mapped by ι_p to a *distinct* geodesic path from $x = \iota_p(y_p)$ to p. Let $q_x \in T_x$ denote the lift to T_x of the endpoint p of the path $\iota_p([p', y_p])$. Since the two lifted paths are not equal we conclude that $p_x \neq q_x$; that is, these lifts are distinct in T_x.

Nonempty intersection. Next consider the case that $p \neq q$ are distinct points in X and $\iota_p(\operatorname{int}(H_p)) \cap \iota_q(\operatorname{int}(H_q)) \neq \emptyset$. Let $x_p \in H_p$ and $y_q \in H_q$ be such that $\iota_p(x_p) = \iota_q(y_q) = x \in X$. As before we can lift the path $\iota_p([x_p, p'])$ to T_x so that x lifts to x'. Let p_x be the lift of the endpoint $p = \iota_p(p')$. We can lift $\iota_q([y_q, q'])$ to T_x so that x lifts to x'. Let q_x be the lift of q' to T_x. Again, $p_x \neq q_x$.

Deriving a contradiction. In either of these two cases (i.e., ι_p restricted to H_p fails to be injective, or if $p \neq q$ but $\iota_p(\operatorname{int}(H_p))$ intersects $\iota_q(\operatorname{int}(H_q))$), the exponential map ι_x can be extended to the set $D(p_x, s_p) \cup D(q_x, s_q) \subset T_x$. In both cases $p_x \neq q_x$, which we will use to derive a contradiction. Let

$$\Sigma_{p,x} = \iota_x^{-1}(\Sigma) \cap \partial D(p_x, s_p)$$

be the set of points in $\partial D(p_x, s_p)$ that map to Σ under ι_x. Let $H'_p \subset T_x$ denote the convex hull of the set $\Sigma_{p,x}$. Let $\Sigma_{q,x}$ denote the set of points on the boundary of $D(q_x, s_q)$ that map to points of Σ under ι_x, and let $H'_q \subset T_x$ be the convex hull of $\Sigma_{q,x}$. We have

$$\iota_x(H'_p) = \iota_p(H_p) \quad \text{and} \quad \iota_x(H'_q) = \iota_q(H_q).$$

We have $\Sigma_{p,x} \subset \partial D(p_x, s_p) \setminus D(q_x, s_q)$ so $D(p_x, s_p)$ is not strictly contained in $D(q_x, s_q)$. Similarly $D(q_x, s_q)$ is not strictly contained in $D(p_x, s_p)$. The disks are not disjoint because they both contain the origin $x' \in T_x$. They are not equal because they have distinct centers. We conclude that the circles $\partial D(p_x, s_p)$ and $\partial D(q_x, s_q)$ intersect in two points. The line between these points separates all points in $\Sigma_{p,x}$ other than the two points from all points in $\Sigma_{q,x}$ other than the

3.6. Delaunay triangulations

two points. Thus if H_p and H_q are 2-dimensional, the interiors of the convex hulls H_p and H_q are disjoint. This contradicts the assumption that x' is a point in common. □

Lemma 3.6.4. *For each Delaunay 1-cell $\beta \subset X$ there is a pair of Delaunay 2-cells E_1 and E_2 such that $\beta \cup E_1 \cup E_2$ contains a neighborhood of the interior of β.*

Proof. Let $p \in X$ be an interior point of the Voronoi 1-cell V_1 dual to β. There are points $\sigma_1, \sigma_2 \in D(p, s_p) \subset T_p$ so that the segment $[\sigma_1, \sigma_2]$ joining them is mapped to β under the exponential map ι_p; that is,

$$\iota_p([\sigma_1, \sigma_2]) = \beta.$$

Now fix p_0 to be the midpoint of β and let $\gamma : (a, b) \to X$, $a < 0 < b$, parameterize the maximal geodesic through p_0 in the direction perpendicular to the direction of β. Let $\gamma(0) = p_0$ and let $\tilde{\gamma} : (a, b) \to T_{p_0}$ be the lift of γ to T_{p_0}, so $\iota_{p_0} \circ \tilde{\gamma} = \gamma$. The Voronoi cell V_1 is the subset of the image under γ of (a, b) consisting of points with exactly two length-minimizing paths to Σ. We first show that both a, b are finite. For $0 \leq t < b$ the segments $\iota_{p_0}([\sigma_1, \tilde{\gamma}(t)])$ and $\iota_{p_0}([\sigma_2, \tilde{\gamma}(t)])$ are length-minimizing paths from $\gamma(t)$ to $\iota_{p_0}(\sigma_i)$. We have

$$d(\gamma(t), \Sigma) = \sqrt{t^2 + (\kappa/2)^2}$$

where κ is the length of β, which is also the length of the line segment $[\sigma_1, \sigma_2] \in T_{p_0}$. The function $d(x, \Sigma)$ is bounded on X. From this it follows that $b < \infty$. The same argument shows that $a > -\infty$. The points $\gamma(a)$ and $\gamma(b)$ have at least one additional length-minimizing path to Σ, for otherwise a and b would not be maximal. This says that they are Voronoi 0-cells. The polygons $\iota_a(H_{\gamma(a)}), \iota_b(H_{\gamma(b)})$ are the dual Delaunay 2-cells. Both of these sets contain the segment β but they lie on opposite sides. It follows that their union together with β contains a neighborhood of the interior of each interior point of β. □

Theorem 3.6.5. *Every point in X is contained in a unique Delaunay cell. Every Delaunay cell is isometric to a polygon inscribed in a disk of radius less than or equal to* $\mathrm{diam}(X)$.

Proof. The geometric description of Delaunay cells follows immediately from their construction. The uniqueness of the Delaunay cell containing a given point is a consequence of Proposition 3.6.3. We consider existence. Consider the subset of points in $X \setminus \Sigma$ which are contained in Delaunay 1-cells and 2-cells. By Lemma 3.6.4 this set is open in $X \setminus \Sigma$. The union of a 2-cell and 1-cells on its boundary is closed in $X \setminus \Sigma$ and there are only finitely many cells; thus this set is closed in $X \setminus \Sigma$. We conclude this union is open and closed and therefore all of $X \setminus \Sigma$. The points of Σ are Delaunay 0-cells. The cells of the Delaunay decomposition are isometric to convex polygons inscribed in disks. □

Delaunay triangulations. The polygonal description of Delaunay decompositions in Theorem 3.6.5 allows us to define *Delaunay triangulations* of translation surfaces.

Definition 3.6.6. Let ω be a translation surface. Any triangulation of the polygonal cells of the Delaunay decomposition into triangles which does not introduce new vertices is called a *Delaunay triangulation* of ω.

Exercise 3.11. *Draw the Delaunay triangulation of your favorite L-shaped translation surface.*

3.6.1. Large diameter surfaces. In preparation for using Delaunay triangulations to prove that Lebesgue (MSV) measure on the stratum of unit-area translation surfaces is finite, we discuss further geometric properties of translation surfaces with *large diameter*. We first introduce the notion of *injectivity radius*, which measures the "thickness" of a surface at a point.

Definition 3.6.7. Let ω be a translation surface, and let X be the underlying Riemann surface. Let $p \in X$ be a regular point (that is, not a zero of ω), and let $p' \in T_p$ be the origin in the tangent space $T_p = T_p X$ to X at p. The *injectivity radius* r_p at p is the supremum of the set of $r > 0$ so that the exponential map ι_p restricted to the Euclidean disc $D(p', r) \subset T_p$, (where the metric on T_p is determined by ω) is an isometric embedding of $D(p', r)$ in X.

Distance to zeros. With notation as in Definition 3.6.7, recall that $\Sigma \subset X$ denotes the set of zeros of ω. For $p \in X$, we write $s_p = d(p, \Sigma)$ for the distance from p to the set of zeros Σ.

Lemma 3.6.8. *Let $p \in X$, and suppose $s_p > r_p$. Then there is a closed geodesic of length $2r_p$ passing through p.*

Proof. We use the exponential map ι_p to construct an immersion of an open disk of radius r_p into X. Since $r_p < s_p$ and by the definition of injectivity radius, the immersion given by ι_p restricted to the closed disc $\overline{D(p', r_p)} \subset T_p$ cannot be an embedding, for then we could find an embedded disk of radius larger than r_p. Thus, there must be two points $z_1, z_2 \in \partial D(p', r_p)$ which map to the same point $\iota_p(z_1) = \iota_p(z_2) = q \in X$. We now construct a parametric curve $\gamma : [0, 2r_p] \to X$ with domain the segment $[0, 2r_p]$. The map γ sends the interval $[0, r_p]$ isometrically to the image under ι_p of the radius from 0 to z_1 so that 0 maps to p and r_p maps to $\iota_p(z_1)$. That is, identifying T_p with $\mathbb{C}$ with p' identified to 0, and writing $z_1 = r_p e^{i\theta_1}$,

$$\gamma(t) = \iota_p(t e^{i\theta_1}), \quad 0 \le t \le r_p.$$

3.6. Delaunay triangulations

On the second half of its domain γ sends $[r_p, 2r_p]$ isometrically to the image of the radius from z_2 to 0 so that r_p maps to q and $2r_p$ maps to p; that is, if $z_2 = r_p e^{i\theta_2}$,

$$\gamma(t) = \iota_p((2r_p - t)e^{i\theta_2}), \quad r_p \leq t \leq 2r_p.$$

By definition $\gamma(0) = \gamma(2r_p) = p$, so γ is a loop which is a geodesic except perhaps at the points 0 and r_p. We will show that the loop is smooth at 0 and r_p and hence is a closed geodesic. To show it is smooth at r_p it is enough to show that the outward pointing normal to the circle $\partial D(p', r_p)$ at z_1 and the inward pointing normal to $\partial D(p', r_p)$ at z_2 get mapped under (the derivative of) ι_p to the same vector. To show that let $U_1 \subset D(p', r_p) \subset T_p$ be a small open disc whose closure contains z_1 and which is tangent at z_1 to the line perpendicular to the outward normal at z_1. Similarly choose a disc U_2 at z_2. Since the images of U_1 and U_2 under ι_p are disjoint, the outward normal at z_1 must map to the inward normal at z_2.

Smoothness. To verify smoothness at 0 we calculate the holonomy around the loop γ. Let θ be the angle the geodesic $[z_2, 0]$ makes with $[0, z_1]$ at 0. Let v and v' be the unit vectors at 0 tangent to these geodesics. If we parallel translate the vector $D\iota(v)$ along the path γ, it returns to $D\iota(-v')$.

Since ω is a translation surface the tangent vectors along any closed path must coincide so $D\iota(v) = D\iota(-v')$ or equivalently $v = -v'$. This shows that the loop γ is actually a closed geodesic of length $2r_p$. $\square$

Lemma 3.6.9. *Let $\gamma : [a, b] \to X$ be a closed geodesic, with the property that there is an $s > 0$ so that for all $t \in [a, b]$,*

$$s_{\gamma(t)} \geq s.$$

Then γ is a waist curve in a cylinder whose height is at least $2s$.

Proof. For each $t \in [a, b]$, consider the line segment of length s with one endpoint $\gamma(t)$, perpendicular to γ, each segment leaving on the same side of γ. Since $s \leq s_{\gamma(t)}$ the segment does not run into a zero. Fix $0 \leq u \leq s$ and denote by $\beta_u(t)$ the point at distance u along this perpendicular segment. Since the metric is Euclidean, the path $\{\beta_u(t); a \leq t \leq b\}$ is a closed geodesic parallel to $\beta_0(t) = \gamma(t)$. This is possible for the perpendiculars in both directions, thus creating a cylinder whose height is at least $2s$. $\square$

Decomposing the surface. We now show that the part of the surface which is further than a certain fixed distance from the zeros can be decomposed into cylinders. We now fix our translation surface ω to have area 1. Set $s_0 = \frac{2}{\sqrt{\pi}}$ and let $E = \{p \in X : s_p \geq s_0\}$.

Theorem 3.6.10. *E is contained in a union of disjoint cylinders.*

Proof. Recall that a cylinder is maximal (Definition 2.4.2) if it is not contained in any larger cylinder. Consider the collection of maximal cylinders with the property that their circumference is less than their height. Two distinct cylinders in this set are disjoint. In fact if C is such a cylinder and $p \in C$, then the unique shortest curve through p is a waist curve of C: any other geodesic through p must cross both boundaries of C and hence have length greater than or equal to the height of C. Thus if p is contained in some other cylinder C' in this collection, then C and C' have a waist curve in common, and since they are maximal, they must coincide. To prove our result it suffices to show that every point in E is contained in such a cylinder. Let $p \in E$. Then $s_p \geq s_0$ by definition. By the definition of the injectivity radius r_p, the surface contains an embedded disk of radius r_p at p. Since the area of the entire surface is 1 we have $\pi r_p^2 < 1$. Since

$$s_p \geq s_0 > \frac{1}{\sqrt{\pi}} > r_p,$$

Lemma 3.6.8 implies that there is a closed geodesic γ of length L through p with $L < \frac{2}{\sqrt{\pi}}$. Let $d = d(\gamma, \Sigma)$ be the distance from γ to the zero set Σ. By Lemma 3.6.9 there is an embedded cylinder around γ with height at least $2d$. The boundary of this cylinder must contain a singularity $q \in \Sigma$. The maximum distance between p and a point on the boundary of the cylinder is $\sqrt{\frac{1}{\pi} + d^2}$. On the other hand, since $p \in E$ the distance from p to q is at least s_p. This gives

$$\sqrt{\frac{1}{\pi} + d^2} \geq s_p > s_0 = \frac{2}{\sqrt{\pi}},$$

which implies $d > \frac{1}{\sqrt{\pi}}$. The height of the cylinder is at least $2d > \frac{2}{\sqrt{\pi}}$ which is greater than the circumference. $\square$

Delaunay edges. Theorem 3.6.10 allows us to give a fairly precise description of edges in Delaunay triangulations.

Proposition 3.6.11. *As above, let $s_0 = \frac{2}{\sqrt{\pi}}$ and let ω be an area 1 translation surface. Any Delaunay triangulation of ω consists of edges e which either*

- *have length $\ell(e) \leq 2s_0$*
- *or crosses a cylinder C whose height $h = h(C)$ is greater than its circumference $r = r(C)$.*

In the second case, if e crosses a cylinder C, then

$$h \leq \ell(e) \leq \sqrt{h^2 + r^2}.$$

Proof. Given a Delaunay edge e, let p be a point in the dual Voronoi 1-cell R_e. We will split the proof into two cases. First assume p is within distance s_0 of the zeros, i.e., $s_p \leq s_0$; that is, $p \notin E = \{p \in X : s_p \geq s_0\}$. Then the Delaunay edge e is the isometric image under ι_p of a secant line in $D(p', s_p)$. Thus

$$\ell(e) \leq 2s_p \leq 2s_0.$$

Now suppose $p \in E$. By Theorem 3.6.9, $p \in C$ for some cylinder C. Let r be the circumference of C. Let R be a rectangle inscribed in $D(p', s_p) \subset T_p$ with vertices on $\partial D(p', s_p)$ so that one pair of opposite sides of R is mapped by ι_p to closed loops of length r parallel to the waist curve of C and the other pair of opposite sides is perpendicular to the waist curve. By rotating ω, we can assume that the waist curve is vertical. Since C does not contain elements of Σ, $r \leq 2s_p$, the diameter of the disc, so this construction is possible.

Any vertical segment in R maps under ι_p to a waist curve in C so its opposite endpoints are identified. Thus the two horizontal sides of R map under ι_p to the same segment in the surface, and so $\iota_p(R)$ is a cylinder. Extend the sides of R which are parallel to the waist curve to infinite lines. Let R' be the set of points in $D(p', s_p)$ which are between these lines. By construction $\iota_p(R') = \iota_p(R)$. In particular no points in the interior of R' map to elements of the zero set Σ. The secant line $\iota_p^{-1}(e) \cap D(p', s_p)$ is a secant line in $D(p', s_p)$ connecting points $z_1, z_2 \in \iota_p^{-1}(\Sigma)$. If these points are contained in the same component of $D(p', s_p) \setminus R'$, then the distance between them is at most the circumference r of the cylinder C, and

$$r \leq s_0 = \frac{2}{\sqrt{\pi}}.$$

If z_1, z_2 are in different components of $D(p', s_p) \setminus R'$, then the secant line crosses R. Thus the length of the edge e is bounded below by the height h of C; i.e.,

$$\ell(e) \geq h.$$

The length $\ell(e)$ is bounded above by the diameter of $D(p', s_p)$, which is equal to the diameter of the inscribed rectangle R. One dimension of R is equal to the circumference r of C; the other is bounded above by the height h of C. This gives the upper bound

$$\ell(e) \leq \sqrt{h^2 + r^2}. \qquad \square$$

3.7. Finiteness of the MSV measure

We conclude this chapter by using the machinery of Delaunay triangulations developed above in §3.6 to show the finiteness of the natural absolutely continuous $SL(2, \mathbb{R})$-invariant measure μ on connected components $\mathcal{H}$ of strata $\mathcal{H}(\alpha)$ of unit-area translation surfaces. This theorem, first shown independently by Veech [162] and Masur [118] for strata of translation surfaces, was shown to hold for all strata of quadratic differentials in [162] (and in the case of the so-called principal stratum in [118]). The proof we now sketch was given in Masur-Smillie [**121**]. We note that there are approaches to proving this using interval exchange transformations (see, for example, the survey of Yoccoz [**174**]) and by carefully studying degenerations (see §6.6 for this approach).

Theorem 3.7.1. *For each connected component $\mathcal{H}$ of a stratum $\mathcal{H}(\alpha)$ of unit-area translation surfaces, we have*

$$\mu(\mathcal{H}) < \infty.$$

Proof. We have shown that associated to each $\omega \in \mathcal{H}$ is a Delaunay triangulation. Up to the action of the mapping class group there are a finite number of possible triangulations $\mathcal{T}$ (see, for example, Harer [**83**]). Fixing a triangulation $\mathcal{T}$, define

$$\mathcal{H}_{\mathcal{T}} = \{\omega \in \mathcal{H} : \mathcal{T} \text{ is a Delaunay triangulation for } \omega\}.$$

It suffices to show that the measure $\mu(\mathcal{H}_{\mathcal{T}})$ is finite. Choose a basis

$$\beta_1, \ldots, \beta_{2g+n-1}$$

for $H_1(S, \Sigma, \mathbb{Z})$ consisting of edges of $\mathcal{T}$. Proposition 3.6.11 implies that if β_i satisfies

$$\ell(\beta_i) = |z_{\beta_i}| = \left|\int_{\beta_i} \omega\right| \geq 2s_0 = \frac{4}{\sqrt{\pi}},$$

then β_i crosses a cylinder C_i whose height is at least s_0. Moreover the component of the holonomy vector of β_i in the direction of the waist curve of C_i is bounded by the circumference of C_i. If β_k is another saddle connection also satisfying $\ell(\beta_k) \geq 2s_0$, then C_i and C_k are distinct cylinders. Now for each such cylinder C_i that occurs because of such a β_i, a saddle connection γ_i on the boundary of C_i is either homologous to one of the β_j or is a linear combination of β_j. Along with β_i we can use it as a basis element of $H_1(S, \Sigma, \mathbb{Z})$. The collection of β_i and γ_i that come from the cylinders C_i can be completed to a basis for $H_1(S, \Sigma, \mathbb{Z})$ by adding edges in $\mathcal{T}$ that do not cross any C_i. Since Area$(\omega) = 1$

$$\text{Area}(C_i) = h_i r_i \leq 1$$

where $h_i = h(C_i)$ and $r_i = r(C_i)$ are the height and circumference of the cylinder C_i. Thus

$$s_0 \leq h_i \leq \frac{1}{r_i},$$

so

$$|\gamma_j| = r_i \leq \frac{1}{s_0}.$$

We can think of h_i as the length of the projection of the vector z_{β_i} to the perpendicular to $z(\gamma_j)$. For each vector z_{γ_j} representing the holonomy vector of the circumference, the holonomy vector z_{β_i} of the curve β_i crossing the cylinder has two components. The component in the direction of γ_j has length at most $|\gamma_j| = r_i$ and the component in the perpendicular direction has length

3.7. Finiteness of the MSV measure

at most the height h_i which is bounded by $\frac{1}{r_i}$. The set of holonomy vectors of the $z(\beta_i), z(\gamma_j)$ lie in a subset of $\mathbb{R}^4$. By Fubini's theorem (see, for example, [**144**, §8]) this set has Lebesgue measure at most

$$\pi r_i^2 r_i \frac{1}{r_i} = \pi r_i^2 \le \frac{\pi}{s_0^2}.$$

The holonomy of all other saddle connections that are in the triangulation and are not saddles crossing the cylinders have length bounded by $2s_0$. We conclude again by Fubini's theorem that the total Lebesgue measure is bounded in terms of the constant s_0. □

Chapter 4

Dynamical Systems and Ergodic Theory

We now move to studying natural families of dynamical systems associated to a translation surface ω, namely straight line flows ϕ_t^θ in a fixed direction θ. We will consider their associated discretizations, known as interval exchange transformations. We will discuss some important basic properties of these maps, from the point of view of both *ergodic theory* and *topological dynamics*. Broadly speaking, ergodic theory is concerned with the properties of *measure-preserving* systems and with the equidistribution properties of trajectories. Topological dynamics is concerned with the *topological properties* of orbits, namely, their density, and if trajectories are not dense, it is concerned with their orbit closures. We will first discuss topological dynamics (§4.1) of our linear flows before turning to the basics of ergodic theory (§4.2). We'll then discuss (§4.3) important low-genus examples in detail which illustrate some of the interesting phenomena that occur for the ergodic theory of these flows that do not occur in the torus case. In particular there are linear flows with dense orbits which are not equidistributed with respect to Lebesgue measure and, indeed, have multiple invariant measures. In §4.4, we'll show how to bound the number of ergodic measures for the linear flow on a translation surface in terms of the genus of the surface, and in §4.5, we'll discuss the associated 1-dimensional dynamical systems known as *interval exchange transformations*.

4.1. Topological dynamics

We start with a translation surface ω and let ϕ_t^θ be the straight line flow in direction θ. By replacing ω by $e^{i(\pi/2-\theta)}\omega$ we can assume the direction is vertical (that is, the flow is in direction $\pi/2$) and remove the parameter θ. We denote the vertical flow by ϕ_t. We first consider these flows from the topological point of view. Let $\Sigma \subset X$ be the set of zeros of ω on the underlying Riemann surface X.

We take the following convention for a point $p \in X$ which is not a zero of ω (that is, $p \in X \setminus \Sigma$). If there is a t_0 such that $\phi_{t_0} p \in \Sigma$, then if $t_0 < 0$, the orbit $\phi_t p$ is not well-defined for $t < t_0$, and similarly, if $t_0 > 0$, the orbit $\phi_t p$ is not well-defined for $t > t_0$. In these cases we say that the ϕ_t-orbit of p is *finite* in the negative (respectively, positive) direction. There can be $t_0 < 0 < t_0'$ such that $\phi_{t_0} p, \phi_{t_0'} p \in \Sigma$; that is, p is on a vertical saddle connection, in which case we say the orbit is finite in both directions. If the orbit is not finite in a direction, we say it is infinite in that direction. We allow an infinite orbit to be periodic.

Minimal domains. We introduce the concept of a *minimal domain*.

Definition 4.1.1. A *minimal domain* D for the flow ϕ_t is a ϕ_t-invariant open subset of X such that for every $p \in D$ the orbit $\{\phi_t p\}_{t \in \mathbb{R}}$ through p is either

- dense in D in both directions, that is,
$$\overline{\{\phi_t p : t \geq 0\}} = \overline{\{\phi_t p : t \leq 0\}} = \overline{D},$$

- or it hits a singularity in one direction and is dense in D in the other direction. That is, there is a $t_0 \neq 0$ such that $\phi_{t_0}(p) \in \Sigma$. If $t_0 > 0$, $\overline{\{\phi_t p : t < 0\}} = \overline{D}$, and if $t_0 < 0$, $\overline{\{\phi_t p : t > 0\}} = \overline{D}$.

If the entire surface X is a minimal domain, we say the flow ϕ_t is *minimal*.

Our first main result on the topological dynamics of flows on translation surfaces is a precise description of possible minimal domains.

Theorem 4.1.2. *Suppose the ϕ_t-orbit through $p \in X$ is infinite in at least one of the two directions and is not periodic. Then p lies in a minimal domain D, and if $D \subsetneq X$ is not the entire surface X, then D is an open subset whose boundary is a union of vertical saddle connections.*

Returns. In order to prove Theorem 4.1.2, we need the following lemma, which states that vertical flow trajectories return to transverse segments. We use ψ to denote the *horizontal* flow. That is, our convention is that $\phi_t = \phi_t^{\pi/2}$, $\psi_t = \phi_t^0$ on ω. Let $p \in X$, and suppose that the ϕ_t-orbit of p is infinite in the positive direction and not periodic. Let $\alpha^+ = \{\phi_t p : t > 0\}$ be the infinite vertical trajectory. Let $\beta = \{\psi_s p : 0 \leq s \leq s_0\}$ be a finite segment of horizontal trajectory through p. The trajectory β is perpendicular to α^+ where they meet at p.

Lemma 4.1.3. *Let p, α^+, and β be as above. The trajectory α^+ returns to β after leaving p. That is, there is a $t_0 > 0$ such that $\phi_{t_0} p \in \beta$.*

Proof. Since there are a finite number of singularities, there are a finite number of vertical trajectories starting at points of β that hit a singularity before crossing β again. By decreasing s_0 if needed, we can shorten β to a subinterval $\beta' = \{\psi_s p : 0 \leq s \leq s_0'\}$, $s_0' \leq s_0$, with one endpoint p and other endpoint $q = \psi_{s_0'} p$ so that no vertical trajectory leaving β' hits a singularity before returning to β. Consider the image of the interval β' under the flow ϕ_t in the positive vertical direction for some fixed time t_0. This process sweeps out rectangles

$$\{\phi_t \psi_s p : 0 \leq s \leq s_0', 0 \leq t \leq t_0\}.$$

As t_0 grows, the area of these rectangles is growing, and since the total area of the surface ω is finite, the interval β' must return and overlap β for a first time t_0. If the orbit α^+ of p itself does, we are done. Otherwise the trajectory leaving q returns to β' and some trajectory leaving a point $q' \in \beta'$ returns to p at time t_0. Consider the interval $\beta'' \subset \beta'$ with endpoints p and q' and apply the previous analysis to it. Flowing in the forward direction β'' must return to β for a first time $t_1 > t_0$. Now either p returns to β at time t_1 and we are done or q' does. But $q' \in \alpha^+$ and so in this case p returns to β at time $t_0 + t_1$. □

Describing minimal domains. We now give the proof of Theorem 4.1.2. Fix $p \in X$, and assume that the vertical trajectory $\alpha^+ = \{\phi_t p : t \geq 0\}$ is infinite, not periodic, and not dense in the entire surface. Then the limit set

$$A = \{y \in X : \exists t_n \to +\infty \text{ such that } \phi_{t_n} p \to y\}$$

of possible limit points of α^+ satisfies $A \subsetneq X$. Let $p_0 \in \partial A \setminus \Sigma$ be a point of A which is not in the interior of A and not in the set of zeros Σ. Let

$$\gamma = \{\phi_t p_0 : t \in \mathbb{R}\}$$

be the vertical line through p_0. Let

$$I = \{\psi_s p_0 : -s_0 < 0 < s_0\}$$

be an open horizontal segment containing p_0. Since p_0 is not in the interior of A and A is closed, we can write $I \setminus (A \cap I)$ as a union of disjoint open horizontal segments I_n; that is,

$$I \setminus (A \cap I) = \bigcup \{\psi_s p_0 : s \in I_n\},$$

where by abuse of notation we are using I_n to denote both the segment and the interval of times. Each endpoint p_n other than possibly $-s_0$ and s_0 of any of these intervals I_n lies in A. By Lemma 4.1.3 the vertical line through p_n must be a saddle connection; otherwise the line would intersect I_n, which would be a contradiction, since A is by construction invariant under the vertical flow ϕ_t.

Since there are finitely many vertical saddle connections, there are only finitely many that intersect I. Therefore there are only finitely many intervals I_n which implies that p_0 is an endpoint of one of them. This says that p_0 is on a saddle connection. We next argue that A has nonempty interior. If A has empty interior, then the above argument shows that A consists of a compact set of vertical saddle connections. Let Γ be this collection and let U be a small neighborhood of Γ. Since A consists of saddle connections, for large enough time, the infinite vertical line α must remain in U. However since it is parallel to each saddle connection in Γ it must stay a constant distance from Γ and hence cannot accumulate on Γ, a contradiction.

Next let $p_0 \in \partial A \cap \Sigma$. Then the vertical trajectory must pass arbitrarily close to p_0 and hence arbitrarily close to vertical segments leaving p_0. Points on such segments are in A and hence by the above arguments lie on a saddle connection. Thus p_0 itself lies on a saddle connection. This concludes the proof of Theorem 4.1.2. $\square$

Minimal directions. Theorem 4.1.2 allows us to conclude that the surface decomposes in each direction into minimal domains and cylinders and in fact that the set of nonminimal directions (that is, directions where the flow is not minimal in the whole surface) is countable.

Corollary 4.1.4. *For any direction θ the surface decomposes into the closure of minimal domains and cylinders in direction θ. There are only countably many directions in which the flow ϕ_t^θ is not minimal in the whole surface.*

Proof. Every point $p \in X$ with an infinite ϕ_t^θ trajectory (in either the positive or negative direction) and that is not periodic belongs to a minimal domain. If it is periodic, it belongs to a cylinder. The set of points for which one of these possibilities holds is open and dense in the surface. This proves the first statement. The second follows from the fact that there are only finitely many zeros and hence only countably many saddle connections (as we showed in Lemma 2.4.7). If θ is a direction such that there is no saddle connection for the flow ϕ_t^θ, then the flow ϕ_t^θ is minimal. $\square$

4.1.1. Strebel directions. Directions without minimal domains are known as *Strebel* directions. In a Strebel direction θ, the surface decomposes into a union of cylinders. In many of our examples, there are obvious Strebel directions.

Exercise 4.1. *Show that the vertical and horizontal directions are Strebel directions on the following genus 2 translation surfaces:*

- *the regular octagon with parallel sides identified by translation and with vertical direction parallel to a side,*
- *the regular decagon with parallel sides identified by translation and with vertical direction parallel to a side,*

- *any L-shaped surface with parallel sides identified by translation and with vertical direction parallel to a side.*

A dichotomy. Recall that for the flat torus $(\mathbb{C}/Z[i], dz)$, rational slopes correspond to Strebel directions, and irrational slopes correspond to minimal directions (Theorem 1.3.1). It is a remarkable fact that in the case of the regular octagon and decagon translation surfaces, every direction is a Strebel direction or a direction in which every line is dense, namely a minimal direction. This dichotomy, due to Veech [164], which holds for a very general family of highly symmetric surfaces known as *lattice* or *Veech* surfaces, will be discussed in Chapter 7.

Multiple minimal domains. For a different example we consider the translation surface obtained from two tori glued along a vertical slit. (An example is given in Figure 2.5.) If the vertical flow is minimal on each torus, then the surface decomposes into two minimal domains in the vertical direction. The common boundary is the pair of slits. On the other hand the vertical lines on one (or possibly both) tori may be closed as in Figure 2.5. In this case there are either one or two cylinders. We will discuss this example further in §4.3.1.

4.2. Ergodic theory

We now turn to the *ergodic* properties of linear flows on translation surfaces, with a particular focus on the equidistribution properties of long trajectories with respect to Lebesgue measure. We start by recalling some basic notions from classical ergodic theory. Excellent references for these and for more background in ergodic theory are Walters's classic book [170] and the more recent book of Einsiedler-Ward [52].

4.2.1. Basic notions of ergodic theory. Suppose μ is a probability measure on a space X, with underlying σ-algebra $\mathcal{B}$. In most of what follows the underlying σ-algebra will be implicit. The example we will be working towards understanding is where our space will be a translation surface and the measure μ will be a measure invariant under a directional flow, with $\mathcal{B}$ being the Borel σ-algebra. We start by recalling the definition of a measure-preserving transformation.

Definition 4.2.1. A transformation $T : X \to X$ is μ-*measure-preserving* if
$$\mu(T^{-1}(E)) = \mu(E)$$
for all μ-measurable sets $E \in \mathcal{B}$. A μ-integrable real-valued function $f \in L^1(X, \mu)$ is *T-invariant* if
$$(f \circ T)(x) = f(x)$$
for μ-almost all $x \in X$.

Definition 4.2.2. A μ-measure-preserving transformation T is *ergodic* if every T-invariant function $f \in L^1(\mu)$ is μ-almost surely constant; that is, there is a $c \in \mathbb{R}$ such that $f(x) = c$ for μ-almost every $x \in X$. Equivalently, T is ergodic if every T-invariant set $E \in \mathcal{B}$ is either μ-null ($\mu(E) = 0$) or μ-conull ($\mu(E^c) = 0$).

Flows. A flow $\phi_t : X \to X$ is μ-measure-preserving if $\mu(\phi_t(E)) = \mu(E)$ for all μ-measurable sets E and all $t \in \mathbb{R}$. In particular, each transformation ϕ_t is μ-measure-preserving as a transformation, since $\mu(\phi_t^{-1}(E)) = \mu(\phi_{-t}(E)) = \mu(E)$.

Definition 4.2.3. A μ-measure-preserving flow ϕ_t is *ergodic* if every flow-invariant $L^1(X,\mu)$-function f (that is, $(f \circ \phi_t)(x) = f(x)$ for all $t \in \mathbb{R}$ and μ-almost all x) is constant almost everywhere.

The Birkhoff Ergodic Theorem. The fundamental theorem of ergodic theory is the *Birkhoff Ergodic Theorem*, which states, informally, that time averages converge to space averages. That is, averaging a function over an orbit of a transformation (or a flow) converges to the average of the function on the space. We state the theorem for flows, since we will mostly be working with flows. For a precise reference, see, for example, Walters [170, §1.6] for transformations and Einsiedler-Ward [52, Corollary 8.15] for flows.

Theorem 4.2.4. *Let ϕ_t be a measure-preserving flow on a probability space (X, μ). Then for any $f \in L^1(X, \mu)$ and for μ-a.e. $x \in X$ the time average of f along the ϕ_t-orbit*

$$f^*(x) = \lim_{T \to \infty} \frac{1}{T} \int_0^T f(\phi_t(x)) dt$$

is well-defined. Moreover, $f^* \in L^1(X,\mu)$ is ϕ_t-invariant and

$$\int_X f^* d\mu = \int_X f d\mu.$$

Ergodicity. If we add the assumption that ϕ_t is ergodic, we have the following corollary (which is also often called the Birkhoff Ergodic Theorem):

Corollary 4.2.5 (Birkhoff Ergodic Theorem). *If in addition the flow ϕ_t is ergodic with respect to μ, then by ϕ_t-invariance, f^* is constant almost everywhere. Since X is a probability space, we have that $f^*(x) = \int_X f d\mu$ for μ-almost every x; that is, the time average is the space average. Precisely, for μ-a.e. $x \in X$,*

$$\lim_{T \to \infty} \frac{1}{T} \int_0^T f(\phi_t(x)) dt = \int_X f d\mu.$$

4.2. Ergodic theory

Generic points. If the phase space X in addition has a *topological structure*, we can restrict the functions we average over orbits (known as *observables*) to *continuous* functions. We now assume X is a compact metric space, μ is a Borel probability measure, and ϕ_t is a measure-preserving flow.

Definition 4.2.6. A point $x \in X$ is μ-*generic* for ϕ_t if for all continuous functions $f \in C(X)$,

$$\lim_{T \to \infty} \frac{1}{T} \int_0^T f(\phi_t(x)) dt = \int_X f d\mu.$$

Unique ergodicity. A key insight in ergodic theory is that if there is a *unique* invariant measure, we can make very strong statements about genericity. Unique ergodicity is a *topological property* with measure-theoretic consequences.

Definition 4.2.7. A flow ϕ_t on a compact metric space X is *uniquely ergodic* if there is a unique ϕ_t-invariant Borel probability measure μ.

We remark that if there is a unique invariant Borel probability measure, it is ergodic. Unique ergodicity can be characterized in terms of genericity:

Theorem 4.2.8. *Let X be a compact metric space, and let ϕ_t be a flow on X. The following are equivalent:*

(1) *ϕ_t is uniquely ergodic.*

(2) *There is a ϕ_t-invariant Borel probability measure μ such that every point is generic for μ.*

Proof. We first show that unique ergodicity implies every point is μ-generic. For $x \in X$ and $t > 0$ define the probability measure δ_x^t to be the Lebesgue probability measure supported on the orbit segment $\{\phi_s x : 0 \le s \le t\}$. That is, for any $f \in C(X)$, we define

$$\int_X f d\delta_x^t = \frac{1}{t} \int_0^t (f \circ \phi_s)(x) ds.$$

We leave as an exercise below (Exercise 4.2) the fact that for any sequence $t_n \to \infty$, there is a subsequence t_{n_k} such that $\delta_x^{t_{n_k}}$ converges in the weak-* topology to an invariant probability measure. Since ϕ_t is uniquely ergodic, this limit measure must be μ. Thus, we have that δ_x^t converges in the weak-* topology to μ for any x, that is, that every x is μ-generic.

To show the converse, suppose ν is an invariant probability measure. We want to show $\mu = \nu$. Let $f \in C(X)$. Since ν is ϕ_s-invariant, for each s,

$$\int_X (f \circ \phi_s)(x) d\nu(x) = \int_X f(x) d\nu(x).$$

Now set
$$F_t(x) = \frac{1}{t}\int_0^t (f \circ \phi_s)(x)ds.$$

Applying Fubini's theorem [**144**, §8] to change the order of integration, we see that for any $t > 0$

$$\int_X f(x)d\nu(x) = \frac{1}{t}\int_0^t \left(\int_X (f \circ \phi_s)(x)d\nu(x)\right)ds = \int_X F_t(x)d\nu(x).$$

Since every point x is μ-generic,

$$\lim_{t\to\infty} F_t(x) = \int_X f(x)d\mu(x).$$

Since f is continuous, there is a constant C such that $\|F_t(x)\|_\infty \leq C$ for all t. Taking the limit as $t \to \infty$ by applying the Dominated Convergence Theorem (see, for example, [**144**, 1.34]) to the sequence $F_t(x)$, we obtain

$$\int_X f(x)d\nu(x) = \lim_{t\to\infty} \int_X F_t(x)d\nu(x)$$
$$= \int_X \left(\int_X f(x)d\mu(x)\right)d\nu(x)$$
$$= \int_X f(x)d\mu(x).$$

The last equality holds since ν is a probability measure. Since this holds for all continuous f we conclude $\nu = \mu$. □

Exercise 4.2. *Fix notation as in Theorem 4.2.8. Using the Banach-Alaoglu theorem (see, for example, [**145**, Theorem 3.15]) prove the statement that for any sequence $t_n \to \infty$ there is a subsequence t_{n_k} such that $\delta_x^{t_{n_k}}$ converges to a ϕ_t-invariant probability measure. The Banach-Aloglu theorem states that the closed unit ball of the dual space of a normed vector space is compact in the weak-* topology. Apply it to the space of probability measures on X.*

Almost sure unique ergodicity. We will see in Theorem 5.1.3 below that almost every translation surface has a uniquely ergodic vertical flow.

4.3. Low genus examples

We discussed in §1.3 the case of the dynamics of linear flows on the torus, and we showed (Exercise 1.2 and Theorem 1.3.1) that the behavior of a linear flow on the square torus $\mathbb{C}/\mathbb{Z}[i]$ is completely determined by the slope: if the slope is

4.3. Low genus examples

rational, every orbit is periodic, and if the slope is irrational, every orbit equidistributes. We leave as an exercise generalizing this to all flat tori:

Exercise 4.3. *Let $\Lambda = g\mathbb{Z}[i] \subset \mathbb{C}$, $g \in SL(2, \mathbb{R})$, be a unimodular lattice, and consider the translation surface structure on the torus $(\mathbb{C}/\Lambda, dz)$. Show that the linear flow ϕ_t^θ has either every orbit periodic (if the slope $\tan \theta$ is contained in the set of slopes of elements of Λ) or every orbit equidistributes (if the slope $\tan \theta$ is not the slope of a lattice vector).*

Lattice surfaces. In Chapter 7 we will show that a class of translation surfaces, known as *lattice surfaces*, that includes the regular octagon (and in fact, all regular 2g-gons) satisfies the same dichotomy as the torus, that for every direction θ either the flow ϕ_t^θ is *Strebel* (that is, the surface decomposes into cylinders in direction θ) or is uniquely ergodic.

4.3.1. Veech nonergodic example.

What makes the study of linear flows on translation surfaces particularly rich is that the behavior discussed above is by far from the only possibility. Once we move past genus 1, we can construct many examples of minimal but *nonuniquely ergodic* linear flows. We describe here a particularly important construction originally due to Veech [**161**], who described it as a $\mathbb{Z}/2\mathbb{Z}$ skew-product of a rotation. Masur-Smillie [**121**] showed how to turn these examples into examples of minimal nonuniquely ergodic linear flows on translation surfaces. Other examples of minimal nonuniquely ergodic linear flows on translation surfaces were constructed by Sataev [**146**], and examples of minimal nonuniquely ergodic interval exchanges transformations (which we will discuss further below in §4.5) were constructed by Keane [**101**]. Our exposition draws on the survey by Masur [**120**].

A skew-product construction. Fix $0 < \alpha < 1$. Consider the product space
$$Y = \mathbb{R}/\mathbb{Z} \times \mathbb{Z}/2\mathbb{Z},$$
and for $\theta \in (0, 1)$ we define the map $T_{\theta, \alpha} : Y \to Y$ by
$$T_{\theta, \alpha}(y, i) = (y + \theta, i + \mathbf{1}_{[0, \alpha)}),$$
where addition in the first coordinate is modulo $\mathbb{Z}$, and in the second is in $\mathbb{Z}/2\mathbb{Z}$. There is a geometric interpretation of this system: Y can be thought of as two copies of the unit circle $S^1 = \{e^{2\pi i t} : t \in [0, 1)\}$. Mark off on each circle the segment
$$J_\alpha = \{e^{2\pi i t} : t \in [0, \alpha)\}$$
of angular width $2\pi\alpha$ in the counterclockwise direction with one endpoint at 0. The map $T_{\theta, \alpha}$ rotates counterclockwise by angle $2\pi\theta$ until the first time the orbit lands in J_α in the circle you started in. Once you land in J_α, switch to the corresponding point in the other circle and then continue rotating by angle $2\pi\theta$ until the orbit lands in J_α, when you switch back to the first circle and so

forth. We let m denote the probability measure on Y given by the *normalized product measure* of Lebesgue measure on $\mathbb{R}/\mathbb{Z}$ and counting measure on $\mathbb{Z}/2\mathbb{Z}$. We leave as an exercise the fact that $T_{\theta,\alpha}$ preserves m.

Exercise 4.4. *Show that $T_{\theta,\alpha}$ preserves the measure m as described above.*

Continued fractions. We will need some basic information about the *continued fraction* expansion of a real number. Fix $x \in (0,1)$, define $a_0(x) = \lfloor \frac{1}{x} \rfloor$, and define the Gauss map $\mathcal{G}$ by

$$\mathcal{G}(x) = \left\{ \frac{1}{x} \right\} = \frac{1}{x} - a_0(x).$$

Define $a_i(x) = a_0(\mathcal{G}^i(x))$. This is well-defined for all integers $i > 0$ for $x \notin \mathbb{Q}$. In this case, we have

$$x = \cfrac{1}{a_0 + \cfrac{1}{a_1 + \cfrac{1}{a_2 + \dots}}}.$$

We say x has *unbounded partial quotients* if the sequence of integers $\{a_n(x)\}_{n \geq 0}$ is unbounded. We refer the interested reader to [105] for more details on continued fractions and their connections to Diophantine approximation.

Unbounded partial quotients. Veech [161] showed that if θ is irrational and has unbounded partial quotients, there are irrational numbers α so that $T_{\theta,\alpha}$ is minimal and such that the Lebesgue measure is not ergodic. In particular the system is not uniquely ergodic because, as remarked earlier, if there is a unique invariant probability measure, then it would automatically have to be ergodic. We will describe how to realize this construction as the first return map for a linear flow on a translation surface.

An associated linear flow. We recall from §2.1.1 (in particular Figure 2.5) a translation surface obtained by gluing two tori along a slit. We describe a slightly more general construction to obtain a family of such genus 2 surfaces. Start with a unit square torus $(\mathbb{C}/\mathbb{Z}[i], dz)$, which we view as the unit square with lower left vertex at $(0,0)$ whose opposite sides are identified. Let σ be a segment in $\mathbb{C}$ joining 0 to $z_0 = x_0 + iy_0$. Take two copies of $(\mathbb{C}/\mathbb{Z}[i], dz)$, each slit along σ, and identify the left side of σ on one copy to the right side in another. As we saw in §2.1.1, this results in a genus 2 translation surface $\omega = \omega_{z_0}$, since the total angle around each of the points 0 and z_0 is 4π, and thus they correspond to two zeros of order 1 of ω. The resulting surface of genus 2 is partitioned into two subsurfaces which we denote M_σ^+ and M_σ^- separated from each other by the union of the two slits.

Vertical slits. In the special case that $z_0 = i\alpha$, the resulting surface has two cylinders in the vertical direction, one for each of the subsurfaces described above. Recall the notion in §4.1 of a first return map of a flow ϕ_t^θ to a transversal(s). We leave the following as an exercise.

4.3. Low genus examples

Exercise 4.5. *Show that the first return map to the union of the pair of core curves of these vertical cylinders for the flow ϕ_t^θ in direction θ is the map $T_{\alpha,\theta}$ described above.*

Billiards with barriers. The surface we describe above in the vertical slit case arises from unfolding rational billiards, as described in §2.1.3. Namely, consider the billiard flow in the polgyon P_α, which is a rectangle of length 1 and width 1/2, with an interior vertical barrier of length $(1 - \alpha)/2$ from the midpoint of a horizontal side.

Exercise 4.6. *Show that the translation surface ω_{P_α} obtained by unfolding P_α is the flat surface ω_{z_0} where $z_0 = i\alpha$ is described in Figure 4.1. See Figure 4.2 for a pictorial guide. The surface one gets is precisely the same as in Figure 2.4. Then recall how we passed from Figure 2.4 to Figure 2.5 by realizing the surface as a pair of tori glued along a slit. Notice that the barrier length $\frac{(1-\alpha)}{2}$ leads to slit length α in Figure 2.5.*

Irrational slits. Returning to our general construction of two tori glued along a slit σ, we say that the slit $\sigma = \sigma(z_0)$ connecting 0 to $z_0 = x_0 + iy_0$ is *irrational* if either

$$x_0, y_0 \neq 0 \quad \text{and} \quad \frac{y_0 + m}{x_0 + n} \notin \mathbb{Q} \quad \text{for all nonnegative integers } m, n$$

or one coordinate is 0 and the other is irrational. Veech's construction started with an irrational rotation θ and showed that there are lengths of gluing intervals α (and hence translation surfaces) such that the flow in direction θ is not uniquely ergodic. The following result heads in the opposite direction, fixing a slit and showing the existence of nonergodic directions.

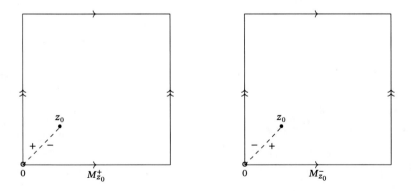

Figure 4.1. Two copies $M_{z_0}^+$ and $M_{z_0}^-$ of the square torus $\mathbb{C}/\mathbb{Z}[i]$ glued along the slit (the dashed line) σ_{z_0} connecting 0 to z_0.

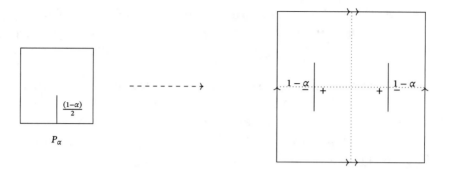

Figure 4.2. Unfolding billiards in the polygon P_α to the translation surface $\omega_{i\alpha}$. The barrier in the polygon P_α starts at the midway point of the bottom side and has height $\frac{(1-\alpha)}{2}$. If the bottom left corner of P_α is at the origin, the top of the barrier is at $\frac{1}{2}(1 + i(1 - \alpha))$. Thus the vertical segments indicated with $\pm$ have length $(1 - \alpha)$, and the associated vertical slit in the decomposition of the surface into two tori has length α.

Theorem 4.3.1. *Suppose the slit $\sigma(z_0)$ is irrational and ω_{z_0} is the corresponding translation surface. Then there exist uncountably many directions θ such that the flow ϕ_t^θ is minimal and not uniquely ergodic.*

Geometric criterion for nonergodicity. To prove Theorem 4.3.1 we first give a geometric criterion theorem, Theorem 4.3.2, of how to find nonergodic linear flows on a translation surface. This criterion was given by Masur-Smillie [**121**]. We fix a translation surface ω with underlying Riemann surface X. Suppose there is a sequence $\{P_n = [A_n, B_n]\}_{n\geq 1}$ of partitions of X into subsurfaces A_n and B_n such that for each $n \geq 1$ the common boundary $\partial A_n = \partial B_n$ of A_n and B_n is a finite union of saddle connections $\{\gamma_{n,i}\}_{i=1}^{k_n}$ of ω all in the same direction θ_n; that is, the holonomy vectors $z_{n,i}$ of the saddle connections satisfy

$$z_{n,i} = \int_{\gamma_{n,i}} \omega = \ell(\gamma_{n,i})e^{i\theta_n}.$$

Here $\ell(\gamma_{n,i})$ is the length. The number of saddle connections k_n is not assumed to be constant.

Limit direction. If we have such a sequence of partitions $\{P_n\}$, then, by passing to a subsequence if needed, we can assume that the directions θ_n converge to a limit θ_∞. Let $\phi_t^{\theta_\infty}$ be the flow in direction θ_∞, and by rotating ω by $e^{i(\pi/2-\theta_\infty)}\omega$ we can assume θ_∞ is the vertical direction. Let

$$h_n = \sum_{i=1}^{k_n} \text{Re}(z_{n,i})$$

4.3. Low genus examples

be the sum of the horizontal components of the holonomy vectors of the saddle connections separating A_n from B_n. Let μ be the Lebesgue measure defined by ω.

Theorem 4.3.2. *With notation as above, suppose there is a partition P_n such that*

   (i) $\lim_{n \to \infty} h_n = 0$,
   (ii) *for some fixed $0 < c \leq c' < 1$, for all n, $c \leq \mu(A_n) \leq c'$, and*
   (iii) $\sum_{n=1}^{\infty} \mu(A_n \Delta A_{n+1}) < \infty$.

Then $\phi_t^{\theta_\infty}$ is not ergodic with respect to the Lebesgue measure μ.

Symmetric differences. We begin the proof of Theorem 4.3.2. We begin with a remark on notation. We use notation $A \Delta B$ to denote the symmetric difference of sets; that is,

$$A \Delta B = (A \cap B^c) \cup (B \cap A^c).$$

We will use both A^c and $X \setminus A$ to denote the complement of a set A.

Borel-Cantelli ideas. To prove Theorem 4.3.2, we define the *liminf* sets A_∞, B_∞ by

$$A_\infty = \liminf A_n = \{x : \exists N \text{ such that for } n \geq N, x \in A_n\},$$
$$B_\infty = \liminf B_n = \{x : \exists N \text{ such that for } n \geq N, x \in B_n\}.$$

The first step in the proof of Theorem 4.3.2 is to show that the sets A_∞ and B_∞ constructed from A_n, B_n satisfy the following four properties:

Properties of the partition.

   Full measure: $\mu(X \setminus (A_\infty \cup B_\infty)) = 0$.
   Disjointness: $A_\infty \cap B_\infty = \emptyset$.
   Approximation: $\mu(A_\infty \Delta A_n) \to 0$ as $n \to \infty$.
   Nontriviality: $0 < \mu(A_\infty) < 1$.

Conclusions. To show our four claimed conclusions, we recall the Borel-Cantelli lemma (see, for example, Feller [66]).

Lemma 4.3.3 (Borel-Cantelli). *If ν is a probability measure on a space S and S_n is a sequence of ν-measurable subsets of S with $\sum \nu(S_n) < \infty$, then*

$$\nu(s \in S : s \in S_n \text{ for infinitely many } n) = 0.$$

We now prove that the four claimed conclusions hold. By hypothesis (iii) in Theorem 4.3.2 and the Borel-Cantelli Lemma 4.3.3, the set of x' which are in infinitely many $A_n \Delta A_{n+1}$ has μ-measure 0. From this we have the claimed full measure property of $A_\infty \cup B_\infty$. The disjointness of A_∞ and B_∞ follows directly by the construction. To see the approximation property, note that

$$A_\infty \Delta A_n \subset \bigcup_{i=n}^{\infty} A_i \Delta A_{i+1}$$

so that

$$\mu(A_\infty \Delta A_n) \leq \sum_{i=n}^{\infty} \mu(A_i \Delta A_{i+1}).$$

By hypothesis (iii) of Theorem 4.3.2 the right-hand side goes to 0 as $n \to \infty$, so $\mu(A_\infty \Delta A_n) \to 0$ as $n \to \infty$. The nontriviality of A_∞ follows from the approximation property and hypothesis (ii) of Theorem 4.3.2.

Almost flow invariance. The next step in the proof of Theorem 4.3.2 is the claim that for any t the set A_∞ is invariant up to a set of μ-measure 0; that is,

$$\mu(\phi_t(A_\infty) \Delta A_\infty) = 0.$$

Suppose to the contrary that for some t_0 we have

$$\mu(\phi_{t_0}(A_\infty) \Delta A_\infty) = \delta > 0.$$

By the approximation property of the partition and hypothesis (i) we may choose n large enough so that

$$\mu(A_\infty \Delta A_n) < \delta/8 \quad \text{and} \quad t_0 h_n < \delta/8.$$

Since ϕ_{t_0} is μ-preserving, the first inequality above and the assumption give

$$\mu(\phi_{t_0}(A_n) \Delta A_n) \geq \delta - 2\delta/8 = 3\delta/4.$$

Thus at time t_0, for each p in a subset $Y_n \subset A_n$ of measure at least $3\delta/8$, the vertical line of length t_0 with endpoint p must intersect the complement of A_n. This set Y_n can be expressed as arising from starting with the set of points on the boundary of A_n and flowing from those points vertically for time at most t_0. We can think of this set as a union of parallelograms with vertical sides of length t_0 and two other sides whose horizontal components have length h_n. The measure of this set is $t_0 h_n < \delta/8$, a contradiction, proving the claim.

Flow invariance. To finish the proof of Theorem 4.3.2 we would like to conclude that A_∞ is actually ϕ_t-invariant. However by the claim it is only almost surely invariant for each time. This issue is fixed by the following. Theorem 4.3.2 is a consequence of the following general ergodic-theoretic lemma, which says that an almost-invariant set is almost an actually invariant set.

Lemma 4.3.4. *Let ϕ_t be a flow on a space X preserving a probability measure μ. Suppose there is a set A such that for every t,*

$$\mu(\phi_t(A) \Delta A) = 0.$$

4.3. Low genus examples

Then there is a set A' invariant under ϕ_t with

$$\mu(A \Delta A') = 0.$$

Proof. Let λ be Lebesgue measure on $\mathbf{R}$. Let

$$A' = \{x : \phi_t(x) \in A \text{ for } \lambda \text{ a.e. } t\}.$$

By definition, A' is ϕ_t-invariant. We will show that

$$\mu(A \Delta A') = 0.$$

Let

$$C_0 = \{(x,t) : x \in A\} \quad \text{and} \quad C_1 = \{(x,t) : x \in \phi_t(A)\}.$$

For every t we have

$$\mu(\{x : (x,t) \in C_0 \Delta C_1\}) = \mu(\phi_t(A) \Delta A) = 0.$$

This implies that $(\mu \times \lambda)(C_0 \Delta C_1) = 0$. By Fubini's theorem [**144**, §8] there is a set X' of full μ-measure so that for all $x \in X'$,

$$\lambda(\{t : (x,t) \in C_0 \Delta C_1\}) = 0.$$

If $x \in X' \cap A$, then $(x,t) \in C_0$ so that

$$\lambda(\{t : (x,t) \notin C_1\}) = 0.$$

Therefore the set $\{t : \phi_t(x) \in A\}$ has full λ-measure so $x \in A'$. If $x \in X' \setminus A$, then the set $\{t : (x,t) \notin C_1\}$ has full λ-measure so that $x \notin A'$. Thus $A \Delta A'$ is contained in the complement of X' so $\mu(A \Delta A') = 0$. This finishes the proof of Lemma 4.3.4 and hence Theorem 4.3.2. □

Proving Theorem 4.3.1. We can now give the proof of Theorem 4.3.1. We will refer to Figure 4.1 and Figure 4.3 in helping to explain the proof. Recall that the segment σ joining 0 to z_0 on the square torus in Figure 4.3 gives rise to the pair of saddle connections in Figure 4.1 that cut ω_{z_0} into two tori, denoted M_σ^+ and M_σ^-. Suppose σ' is another geodesic on the torus (Figure 4.3) with the same endpoints as σ. We can take σ' as corresponding to a segment joining 0 to $z_0 + \zeta$ where $\zeta \in \mathbb{Z}[i]$. We can then think of σ' as giving rise to a pair of saddle connections on ω_{z_0}. Namely each time σ' crosses σ it switches from M_σ^+ to M_σ^- or vice versa. This leads to one saddle connection on ω_{z_0}. The map of ω_{z_0} that exchanges M_σ^+ and M_σ^- takes that saddle connection to another saddle connection with the same endpoints and is otherwise disjoint from the first. The pair of saddle connections cut ω_{z_0} into new two tori which we denote $M_{\sigma'}^\pm$.

Symmetric differences. Suppose the saddle connections σ and σ' intersect an odd number of times in their interior so they divide each other into an even number of pieces (as in Figure 4.3). Equivalently, σ and σ' are homologous mod 2. With a slight abuse of notation we will refer to the saddle connections on ω_{z_0} as σ and σ'. The symmetric difference of the decompositions associated to σ and σ' is

$$(M_\sigma^+ \cap M_{\sigma'}^-) \cup (M_\sigma^- \cap M_{\sigma'}^+)$$

and is a union of an even number of parallelograms with sides on σ and σ'. Thus

(4.3.1) $$\operatorname{area}\left((M_\sigma^+ \cap M_{\sigma'}^-) \cup (M_\sigma^- \cap M_{\sigma'}^+)\right) \leq 2|z_\sigma \wedge z_{\sigma'}|,$$

where

(4.3.2) $$z_\sigma = \int_\sigma \omega_{z_0}, \quad z_{\sigma'} = \int_{\sigma'} \omega_{z_0}$$

are the associated holonomy vectors, and we recall that for two complex numbers z, w, $|z \wedge w|$ denotes the area of the parallelogram spanned by z, w.

Our goal is to find an uncountable sequence of directions each of which satisfies the hypotheses of Theorem 4.3.2. Applying that theorem will produce the desired nonuniquely ergodic directions. We will find these sequence of directions by building a tree.

A binary tree. Fix a sequence of positive numbers $\{\rho_j\}_{j \geq 0} \subset \mathbb{R}^+$ with

(4.3.3) $$\sum_{j=0}^\infty \rho_j < \infty.$$

Associated to this sequence, we will build an infinite directed binary tree $\mathcal{T}$, together with a labeling of the vertices of $\mathcal{T}$ by Gaussian integers, that is, a map

$$L : V(\mathcal{T}) \to \mathbb{Z}[i]$$

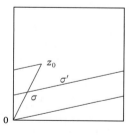

Figure 4.3. The square torus $\mathbb{C}/\mathbb{Z}[i]$ with slits σ (in red) and σ' (in blue) connecting 0 to z_0.

4.3. Low genus examples

from the vertex set $V(\mathcal{T})$ of $\mathcal{T}$ to $\mathbb{Z}[i]$. We denote the jth level of the tree by S_j, with S_0 denoting the root, so there are 2^j vertices in S_j, at distance j from S_0.

Labeling and slits. Each vertex $v \in V(\mathcal{T})$ will be labeled by a Gaussian integer
$$L_v = p_v + q_v i \in \mathbb{Z}[i]$$
which we think of as determining a vector on the torus with endpoints 0 and $z_0 + L_v$. By our irrationality assumption the inverse slope $s_v = \frac{x_0 + p_v}{y_0 + q_v}$ of our segment connecting 0 to $z_0 + L_v$ is an irrational number, $s_v \notin \mathbb{Q}$.

Building the labeling. We will describe how to build the labeling
$$L : V(\mathcal{T}) \to \mathbb{Z}[i]$$
inductively by level. The root S_0 is defined by z_0. Suppose we have defined L on the jth level S_j. Let δ_j be the minimum distance between any two distinct inverse slopes s_v of $z_0 + L_v$ as v varies in S_j. That is,
$$\delta_j = \min\{|s_v - s_u| : v \neq u, v, u \in S_j\}.$$
For each vertex $v \in S_j$ at level j, which we call a *parent*, we will label its *children* $v', v'' \in S_{j+1}$ by
$$L_{v'} = p' + iq', \quad L_{v''} = p'' + iq'' \in \mathbb{Z}[i].$$
To determine the labelings $L_{v'}, L_{v''}$ of the children of v, begin by choosing an even integer $d_j \in 2\mathbb{Z}$ such that

(4.3.4) $$\frac{\rho_j}{(q_v + y_0)(q_v + y_0 + d_j)} < \delta_j/4.$$

Consequences of irrationality. Now consider the inequality

(4.3.5) $$d_j|(L_v + z_0) \wedge \zeta| < \rho_j.$$

The irrationality of the inverse slope $s_v = \frac{x_0 + p_v}{y_0 + q_v}$ implies there are infinitely many solutions $\zeta = m + ni \in \mathbb{Z}[i]$, with $\gcd(m, n) = 1$ of the above inequality. Choose *any* two of them
$$\zeta' = m' + n'i \quad \text{and} \quad \zeta'' = m'' + n''i$$
and define the labels of the children by
$$L_{v'} = L_v + d_j \zeta' = p' + q'i, \quad L_{v''} = L_v + d_j \zeta'' = p'' + q''i,$$
where
$$p' = p_v + d_j m', \quad q' = q_v + d_j n', \quad p'' = p_v + d_j m'', \quad q'' = q_v + d_j n''.$$

Inverse slope distance bounds. We have chosen our labels on children so that the inequalities (4.3.4) and (4.3.5) yield

(4.3.6) $$\delta'_j = |s_v - s_{v'}| \leq \delta_j/4,$$
$$\delta''_j = |s_v - s_{v''}| \leq \delta_j/4.$$

That is to say, the distance between the inverse slope $s_v = \frac{p_v + x_0}{q_v + y_0}$ of the parent slit and the inverse slopes

$$s_{v'} = \frac{p' + x_0}{q' + y_0}, \quad s_{v''} = \frac{p'' + x_0}{q'' + y_0}$$

of its children is bounded by $\delta_j/4$. Thus the distance $|s_{v'} - s_{v''}|$ between the inverse slopes of the two children of v is at most $\delta_j/2$, which implies that the minimal distance δ_{j+1} between pairs at level $(j + 1)$ satisfies

$$\delta_{j+1} < \delta_j/2.$$

Let $\sigma = \sigma_v$ be the segment associated to the parent v and let $\sigma' = \sigma_{v'}$ be the segment associated to the child v'. Since d_j is even, σ' is homologous to σ mod 2. In fact $\sigma' - \sigma$ represents d_j times the primitive class of $\zeta' = m' + n'i$.

Applying Theorem 4.3.2. We want to now apply Theorem 4.3.2. We need to check the three hypotheses. Any infinite geodesic path

$$\{v_j\}_{j \geq 1}, \quad v_j \in S_j, \quad v_{j+1} \text{ a child of } v_j$$

in the tree yields a sequence of partitions of the surface. Hypothesis (ii) holds since each choice of slit gives a partition of the surface into pieces of equal areas. The fact that $\sum_{j=1}^{\infty} \delta_j < \infty$ and equation (4.3.6) imply that the sequence of inverse slopes $\{s_{v_j}\}$ is a Cauchy sequence; hence the directions converge to some θ_{∞}. As we have seen by (4.3.1) the area of the sheet interchange is bounded by $2|v \wedge v'|$, and by (4.3.5) this is bounded by

$$d_j|v \wedge v'| < \rho_j.$$

Now applying (4.3.3) we have that hypothesis (iii) holds. To check hypothesis (i) a calculation shows that $|v||s_{v'} - s_v| \to 0$ and this quantity records the projection of v on the direction perpendicular to v'. From this, (i) follows as well. Thus by Theorem 4.3.2 the flow $\phi^t_{\theta_{\infty}}$ in this limiting direction is not ergodic.

Uncountably many directions. We now show that, in fact, there are uncountably many limiting directions. Since there are only countably many directions that are not minimal, we then can conclude that there are uncountably many limiting directions which are minimal and not uniquely ergodic. Since the set of geodesic paths in the binary tree can be bijectively mapped to $[0, 1]$ (via the expansion of each number by binary digits) there are uncountably many such paths. We therefore need to show that any two path limits are distinct. Suppose

$$r_j = (p_j + x_0)/(q_j + y_0) \to \theta$$

and a distinct sequence
$$r'_j = (p'_j + x_0)/(q'_j + y_0) \to \theta'.$$
Suppose the sequences differ for the first time at stage j, so that the inverse slopes r_j and r'_j satisfy
$$|r_j - r'_j| \geq \delta_j.$$
Then we have
$$|\theta - r_j| \leq \sum_{k \geq j} \delta'_k < \sum_{k \geq j} \frac{\delta_k}{4} < \frac{\delta_j}{2},$$
and similarly $|\theta' - r'_j| < \delta_j/2$, so $\theta \neq \theta'$. This finishes the proof of Theorem 4.3.1. □

4.3.2. Further results. We outline some further results on the ergodic properties of linear flows on translation surfaces. All of this further work was inspired by the existence of these minimal nonuniquely ergodic flow directions and addresses the question of how common these directions are.

Measure of the set of exceptional directions. Theorem 4.3.1 gives uncountably many minimal nonuniquely ergodic flow directions on these surfaces for particular choices of α. We will see in Chapter 5 a deep connection between the $SL(2, \mathbb{R})$-orbits of surfaces and the ergodic properties of their linear flows. An important example is the result of Kerckhoff-Masur-Smillie [103] (which we state precisely as Theorem 5.4.7 in Chapter 5 below) that for any translation surface the set of directions $\theta \in [0, 2\pi)$ such that the flow ϕ_t^θ is not uniquely ergodic has Lebesgue measure 0. So while the set of directions we construct in Theorem 4.3.1 is uncountable, it has Lebesgue measure 0. This opens the question of other notions of size for this exceptional set of directions.

Hausdorff dimension. The family of surfaces with a vertical slit connecting 0 to $i\alpha$ has been extensively studied, as it provides an important source of examples and interesting phenomena, as well as its connection to billiards. Results of Cheung [36] and Cheung-Hubert-Masur [37] give explicit Diophantine conditions on α so that the set of θ with minimal, nonuniquely ergodic linear flow has *Hausdorff dimension* $\frac{1}{2}$. Hausdorff dimension is an important notion of size of a set; we direct the interested reader to, for example, Falconer [63, §2.2]. Note that countable sets have 0 Hausdorff dimension. See §5.5 for further discussion in this direction.

4.4. Number of ergodic measures

We just discussed in §4.3.2 that on every surface and in almost every direction the linear flow has only one invariant measure (namely Lebesgue). We now turn to the question of how many ergodic invariant measures a linear flow on a translation surface may have.

The Choquet simplex. We first recall some general facts about dynamical systems. Note that if $T : X \to X$ is a transformation of a compact metric space and if μ_0 and μ_1 are T-invariant Borel probability measures, then any convex combination

$$\mu_\lambda = \lambda \mu_1 + (1-\lambda)\mu_0, \quad 0 \le \lambda \le 1,$$

is also a T-invariant probability measure (and the same statement holds for flows and, in fact, for general group actions). This observation forms the basis of the following.

Black Box 4.4.1 (Choquet simplex [52, §4.2]). *Let G be a group acting on a topological space X (in most of our examples, G will be $\mathbb{Z}$ or $\mathbb{R}$). The set of G-invariant Borel probability measures*

$$M_G(X) = \{\nu \in M(X) : \nu \text{ is } G\text{-invariant}\}$$

is a convex subset of the set $M(X)$ of probability measures on X, and the extreme points of $M_G(X)$ are the ergodic G-invariant measures. We say a measure is extreme if it cannot be nontrivially written as a linear combination of other measures. We refer to the simplex $M_G(X)$ as the Choquet simplex *for the dynamical system.*

Singularity. To see that the ergodic measures are extreme points, we recall the notion of *singularity* of measures.

Definition 4.4.2. Two probability measures ν_1, ν_2 on a measure space $(X, \mathcal{B})$ are *mutually singular* or simply *singular* with respect to each other if there exists a set $E \in \mathcal{B}$ with

$$\nu_1(E) = 1 \quad \text{and} \quad \nu_2(E) = 0.$$

Exercise 4.7. *Show that if $\phi_t : X \to X$ is a flow on a metric space X and if ν_1 and ν_2 are distinct ergodic invariant measures for ϕ_t, then ν_1 and ν_2 are mutually singular.*

4.4.1. Transverse measures. We first describe the notion of a *transverse-invariant measure* ν associated to a translation surface flow ϕ_t and invariant measure μ for ϕ_t. The measure ν is defined on the set of all arcs transverse to ϕ_t (that is, the arc is not tangent to the vertical direction at any point). Given such an arc $J \subset X$, define

$$(4.4.1) \qquad \nu(J) := \lim_{t \to 0} \frac{1}{t} \mu(\{\phi_s(x) : x \in J, 0 \le s \le t\}).$$

That is, we flow J up to time t, take the μ-measure of the resulting set, normalize by t, and take the limit as $t \to 0$. Note that if J is transverse to the flow ϕ_t, so is the arc $\phi_s J$ for any $s \in \mathbb{R}$.

4.4. Number of ergodic measures

Exercise 4.8. *Show that the measure $\nu(J)$ is well-defined and that it is flow-invariant; that is, for any $s \in \mathbb{R}$,*

$$\nu(\phi_s J) = \nu(J).$$

Straight line segments. If J, for example, is a straight line segment in the Euclidean coordinates defined by ω and if one flows time t, the result is a parallelogram P and one has $\mu(P) = t\nu(J)$. We have the following observation, which we record as a lemma.

Lemma 4.4.3. *The transverse measure ν determines the invariant measure μ.*

Katok's bound. We now state the main theorem of this section, which was proved by Katok [96]. The authors are grateful to Carlos Matheus Santos for very helpful discussions on this section.

Theorem 4.4.4 ([96]). *Suppose ω is a translation surface of genus g and the linear flow ϕ_t^θ in direction θ is minimal. Then there are at most g ergodic invariant probability measures for ϕ_t^θ.*

The flux. To prove this theorem, we will need ingredients from surface topology and symplectic geometry. We start by rotating ω if needed, so we can assume the flow ϕ_t^θ is in the vertical direction. We now write ϕ_t for the flow. Let $\mathcal{M}_1(\phi_t)$ denote the Choquet simplex of invariant probability measures for ϕ_t. Let $\mathcal{M}(\phi_t)$ be the set of all invariant measures. For any $\mu \in \mathcal{M}(\phi_t)$ we now define the *flux* $\psi(\mu) \in H^1(X, \mathbb{R})$, an element of cohomology, that is, a linear functional on homology $H_1(X, \mathbb{R})$. For any closed curve γ, we define

$$\psi(\mu)(\gamma) = \nu(\gamma) = \lim_{t \to 0} \frac{1}{t} \mu(\{\phi_s(x) : x \in \gamma, 0 \leq s \leq t\})$$

to be the transverse measure $\nu(\gamma)$. Note that since ϕ_t is minimal, any closed curve is transverse to ϕ_t.

A map to cohomology. This gives a map

$$\psi : \mathcal{M}(\phi_t) \to H^1(X, \mathbb{R}).$$

We first claim that ψ is injective. Suppose $\mu_1, \mu_2 \in \mathcal{M}(\phi_t)$ and

$$\psi(\mu_1) = \psi(\mu_2).$$

Let γ be any closed curve transverse to the flow. Let p some point in γ. Let

$$b = \psi(\mu_1)(\gamma) = \psi(\mu_2)(\gamma).$$

By abuse of notation, let $\gamma : [0, b] \to X$ be a parametrization of γ by $[0, b]$ with

$$\gamma(0) = \gamma(b) = p.$$

Define a function $G : [0, b] \to \mathbb{R}$ by $G(0) = 0$ and

(4.4.2) $$G(x) = \nu_1[0, x] - \nu_2[0, x],$$

where ν_i is the transverse measure associated to μ_i, $i = 1, 2$.

Exercise 4.9. *Show that the minimality of ϕ_t implies that the measures μ_i and thus the measures ν_i do not have atoms and, so, show that the function G defined in (4.4.2) is continuous.*

First returns. Now let $x \in \gamma$ be any point and let y be the first return to γ of the flow ϕ_t through x; that is,

$$y = \phi_{t_0}(x), \quad t_0 = \min\{t > 0 : \phi_t(x) \in \gamma\}.$$

Closing up the flow line

$$\{\phi_t(x) : 0 \le t \le t_0\}$$

with the segment $[x, y]$ on γ we get a closed curve α and

$$\nu_1([x, y]) = \nu_1(\alpha) = \nu_2(\alpha) = \nu_2([x, y]).$$

We conclude that the function G is invariant under the first return map $T : \gamma \to \gamma$ to γ given by

$$T(q) = \phi_{t_0}(q), \quad t_0 = \min\{t > 0 : \phi_t(q) \in \gamma\}.$$

Since ϕ_t is minimal the orbit of any point $p \in \gamma$ under T is dense in γ, from which it follows that G is constant (since it is continuous and constant on a dense set) and so $\nu_1 = \nu_2$ on γ. We remarked earlier that the transverse measure determines the measure. We conclude, by Lemma 4.4.3 that since $\nu_1 = \nu_2$ we have $\mu_1 = \mu_2$.

Lagrangian subspaces. The cohomology group $H^1(M, \mathbb{R})$ is a 2g-real-dimensional vector space equipped with a *symplectic* bilinear form $i(\cdot, \cdot)$ which is dual to the intersection pairing on homology. For details on symplectic vector spaces, see, for example, [1, §3.1]. We will show that the image of ψ lies in a *Lagrangian subspace*. Recall that a subspace W of a symplectic vector space $(V, \langle \cdot, \cdot, \rangle)$ is called *Lagrangian* if for all $w_1, w_2 \in W$,

$$\langle w_1, w_2 \rangle = 0.$$

We will show the following:

Lemma 4.4.5. *For any $\mu_1, \mu_2 \in \mathcal{M}(\phi_t)$,*

(4.4.3) $$i(\psi(\mu_1), \psi(\mu_2)) = 0.$$

We can then apply the following:

Black Box 4.4.6 ([1, Proposition 5.3.3]). *The dimension of a (real) Lagrangian subspace W of a 2g-dimensional (real) symplectic vector space V is g.*

4.4. Number of ergodic measures

Using Black Box 4.4.6, we can conclude from (4.4.3) that the dimension of $\mathcal{M}(\phi_t)$ is at most g, and since the lines in $\mathcal{M}(\phi_t)$ that correspond to the extreme points of the Choquet simplex $\mathcal{M}_1(\phi_t)$ are the ergodic probability measures, there can be at most g distinct ergodic ϕ_t-invariant probability measures.

The Schwartzmann asymptotic cycle. To show (4.4.3), suppose $\mu \in \mathcal{M}(\phi_t)$ is any ergodic ϕ_t-invariant probability measure. We show how to define the *Schwartzmann asymptotic cycle* associated to μ. This is an element $\lambda_\mu \in H_1(X, \mathbb{R})$ of homology, which will be the Poincaré dual to the cohomology class $\psi(\mu) \in H^1(X, \mathbb{R})$. Let p_0 be a generic point for μ, and let I be a horizontal segment through p_0 and

$$\alpha^+ = \{\phi_t(p_0) : t \in \mathbb{R}\}$$

the ϕ_t-orbit of p_0, that is, the infinite vertical line through p. We saw in Lemma 4.1.3 that α^+ must return infinitely often to I. Fixing $t_0 = 0$, define the sequence t_n inductively by

(4.4.4) $$t_n = \min\{t > t_{n-1} : \phi_t(p_0) \in I\}.$$

Define the sequence of finite segments

$$\alpha_n^+ = \{\phi_t(p_0) : 0 \le t \le t_n\}$$

of α^+ joining p_0 to

$$p_n = \phi_{t_n}(p_0) \in I.$$

Since p_0 is generic, as $n \to \infty$, for all continuous $f : X \to \mathbb{R}$, we have

(4.4.5) $$\lim_{n \to \infty} \frac{1}{t_n} \int_0^{t_n} f(\phi_t(p_0)) dt = \int_X f(x) d\mu(x).$$

Connect p_0 and p_n by a segment $I_n \subset I$ and let γ_n be the closed curve formed by following α_n^+ and then I_n. Equation (4.4.5) implies that the normalized sequence of curves γ_n/t_n converges to a limit in $H_1(X, \mathbb{R})$, which we denote λ_μ. Note that since $|I_n| \le |I|$,

$$\frac{\ell(\gamma_n)}{t_n} \to 1 \quad \text{as } n \to \infty.$$

We now show that the Schwartzmann asymptotic cycles of distinct ergodic invariant measures do not intersect:

Lemma 4.4.7. *Suppose μ_1, μ_2 are distinct ϕ_t-invariant ergodic measures. Then*

$$i(\lambda_{\mu_1}, \lambda_{\mu_2}) = 0.$$

Proof. Choose disjoint horizontal intervals I_1, I_2 and generic points $p_0^i \in I_i$ for ν_i, $i = 1, 2$. Form the closed curves γ_n^1, γ_n^2 as discussed above for the two measures. For sufficiently large n,

$$\frac{\#(\gamma_n^1 \cap \gamma_n^2)}{|\gamma_n^1||\gamma_n^2|} < \frac{2\nu_2(I_1)}{|\gamma_n^1|} + \frac{2\nu_1(I_2)}{|\gamma_n^2|}.$$

The terms on the right go to 0 with n. This shows the intersection number of the asymptotic cycles is 0. The proof is finished by noting that the Schwartzmann cycle $\lambda_\nu \in H_1(X, \mathbb{R})$ for ν is dual to the flux $\psi(\nu) \in H^1(X, \mathbb{R})$. We leave the proof of this statement as Exercise 4.10 below. This concludes the proof of Lemma 4.4.7. □

Exercise 4.10. *Let ν be an ergodic ϕ_t-invariant measure. Show that the Schwartzmann asymptotic cycle $\lambda_\nu \in H_1(M, \mathbb{R})$ and the flux $\psi(\nu) \in H^1(M, \mathbb{R})$ are dual to each other.*

4.5. Interval exchange transformations

We have used the construction of the *first return map* of a linear flow on a translation surface in several discussions above. These first return maps belong to an important class of 1-dimensional dynamical systems known as *interval exchange transformations*, or IETs. We refer the interested reader to Yoccoz's notes [**174**] for an excellent exposition of the details of IETs.

Definition 4.5.1. An *interval exchange transformation*, or *IET*, is a map defined by a vector

$$\lambda = (x_1, x_2, \ldots, x_d) \in \mathbb{R}_+^d$$

and a permutation $\pi \in S_d$ on d letters. Let $|x| = \sum_{i=1}^d x_i$. To define the IET $T_{x,\pi}$, we first define d consecutive subintervals of the interval $I = [0, |x|)$:

$$I_1 = [0, x_1), \quad I_2 = [x_1, x_1 + x_2), \ldots, \quad I_d = [x_1 + \cdots + x_{d-1}, x_1 + \cdots + x_{d-1} + x_d).$$

We define

$$T_{x,\pi} : I \to I$$

which exchanges the intervals I_i according to the permutation π. That is, if $t \in I_j$, then

$$(4.5.1) \qquad T_{x,\pi}(t) = t - \sum_{k<j} x_k + \sum_{\pi(k')<\pi(j)} x_{k'}.$$

Irreducibility. We say an IET is *irreducible* if the underlying permutation π does not preserve any nontrivial subset of $\{1, \ldots, d\}$ of the form $\{1, \ldots, k\}, k < d$. We will only consider irreducible IETS in our discussions.

4.5. Interval exchange transformations

IETs and rotations. The first nontrivial examples of IETs are IETs of two intervals; these turn out to be rotations of the circle:

Exercise 4.11. *Show that a nontrivial IET of two intervals can be seen as a rotation of the circle. Namely, if we take the permutation $\pi = (12)$ and the lengths of the intervals to be $0 < x_1 < 1$, $x_2 = 1 - x_1$, then the resulting map is equivalent to the map*

$$T(y) = y + x_2 \mod 1.$$

3-IETs. IETs of 3 intervals are also closely related to rotations. The next exercise shows that these are all first return maps of rotations to subintervals.

Exercise 4.12. *Show that for any irreducible IET T of 3 intervals on $[0, 1)$ there are α, β so that T can be expressed as the first return map of the map $R_\alpha(y) = y + \alpha \pmod 1$ to the subinterval $[0, \beta)$.*

First returns. If we start with an IET T on an interval I and fix a subinterval J, the first return map is again an IET. We leave the proof as the following:

Exercise 4.13. *If T is an IET of d intervals on an interval I and if J is a subinterval, show that the first return map T_J of T to J is an IET of at most $d + 2$ intervals $J_1, \ldots, J_s$, where $J = \bigcup_{i=1}^{s} J_i$. Furthermore, show that there exist positive integers k_i such that for $i = 1, \ldots, s$,*

$$T_J|J_i = T^{k_i}.$$

Invariant measures. As we can see from (4.5.1), the map $T_{x,\pi}$ is an orientation-preserving piecewise isometry of the interval $I = [0, |x|)$, and thus, Lebesgue measure on I is invariant under the action of T. There are possible discontinuities of T at the endpoints of the intervals I_j, $1 \leq j \leq d$. Especially when studying ergodic or dynamical properties of $T_{x,\pi}$ it will be convenient to normalize so $|x| = 1$ and so $I = [0, 1)$ is the unit interval. The following lemma due to Katok [**97**, Lemma 1] says that any invariant measure for an IET $T = T_{x,\pi}$ can be identified with another IET S (and if the invariant measure is Lebesgue, $S = T$).

Lemma 4.5.2 (Katok [**97**, Lemma 1]). *Suppose μ is an invariant nonatomic Borel probability measure for an IET $T = T_{x,\pi}$ on d intervals, with $|x| = 1$. Then there are an interval exchange map $S : [0, 1) \to [0, 1)$ and a strictly monotone continuous surjective map $\Phi : [0, 1) \to [0, 1)$ which conjugates T to S; that is,*

$$\Phi \circ T = S \circ \Phi$$

and Φ is a measurable isomorphism from μ to Lebesgue measure m; that is, for any Borel measurable $A \subset I$,

$$m(\Phi^{-1}A) = \mu(A).$$

Proof. Define $\Phi : [0, 1) \to [0, 1)$ by the formula
$$\Phi(x) = \mu([0, x]).$$
Then Φ is monotone, continuous, surjective and $\Phi_* \mu = m$, so Φ is an isomorphism between the measure spaces $([0, 1), \mu)$ and $([0, 1), m)$. Define $S : [0, 1) \to [0, 1)$ by the formula $S(x) = \Phi(T(y))$ where $y \in \Phi^{-1}(x)$. In other words, $S = \Phi \circ T \circ \Phi^{-1}$. We show $\Phi^{-1}(x)$ is a point. The other possibility is that it is an interval. We show how to rule this out. We have
$$m(\Phi \circ T \circ \Phi^{-1}(x)) = \mu(T \circ \Phi^{-1}(x)) = \mu(\Phi^{-1}(x)) = m(\{x\}) = 0$$
which says $\Phi \circ T \circ \Phi^{-1}(x)$ is a point. Thus S is well-defined. Now S preserves Lebesgue measure since $\Phi_* \mu = m$. In addition, S is injective, orientation preserving, and continuous except for at most d points. We leave it as an exercise to show that any map with these properties must be an interval exchange of at most d intervals. □

Exercise 4.14. *Suppose $S : [0, 1] \to [0, 1]$ preserves Lebesgue measure and is injective, orientation preserving, and continuous except for at most d points. Show that S is an interval exchange of at most d intervals.*

The space of IETs. For a fixed permutation $\pi \in S_d$ on d letters the set of interval exchange transformations defined on the unit interval $[0, 1)$ with permutation π can be parameterized by the simplex
$$\Delta = \{x \in \mathbb{R}^d_+ : |x| = 1\},$$
which carries a natural $(d-1)$-dimensional Lebesgue measure.

IETs and surface flows. We now give the relationship between straight line flows on translation surfaces ω and IETs, namely, that IETs arise as first return maps of linear flows to transversals. Without loss of generality we assume the straight line flow ϕ_t is in the vertical direction.

Exercise 4.15. *Let J be a transverse interval to the flow ϕ_t. Show that the first return map to J is an IET.*

Choosing a transversal. There are particular choices of J in Exercise 4.15 which will give IETs with a minimal number of exchanged intervals. Choose a horizontal segment I with one endpoint a zero of ω and the other endpoint having the property that the vertical line through that point hits a zero in one of the two directions before returning to I. If ϕ_t is minimal, then every line hits I. The ϕ_t-orbit (that is, the vertical line) through an endpoint of an interval I_i hits a zero or the endpoint of I before returning to I.

Exercise 4.16. *Show that the choice of I as above results in an IET with $d = 2g + k - 1$ intervals, where g is the genus of the underlying surface X and k is the number of zeros of ω.*

4.5. Interval exchange transformations

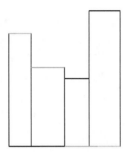

Figure 4.4. A family of genus 2 zippered rectangles. The underlying IET is an exchange of 4-intervals with permutation (4321). The gluings of horizontal sides are indicated by the colors, and the vertical sides are *zipped up* to certain heights and *zipped down* from the bottom after applying the IET.

Zippered rectangles. The choice of I induces a decomposition of the surface into rectangles. The two horizontal sides of each rectangle lie along I, one of which is the subinterval of I defining the IET and the other is the return of that interval by the IET. The vertical sides are either the vertical sides through the endpoints or vertical lines that hit a zero before returning. This picture is often called a *zippered rectangle*; see Figure 4.4. The notion of zippered rectangle was introduced by Veech [**162**].

Suspending an IET. Conversely, an IET together with a piecewise constant positive function on the interval I which is constant on the component intervals I_j can be used to build a closed translation surface (our polygons will be this collection of rectangles), and the gluings are determined by the IET and choices of *heights* of zippers. We will not discuss this construction here but instead refer the interested reader to Yoccoz's detailed notes [**174**].

Hyperelliptic strata. As you showed in Exercise 4.16, the number of intervals for the IET (and indeed the permutation) depends on the genus and the number of zeros of ω. For example, if $\omega \in \mathcal{H}_{\mathrm{hyp}}(2g)$, there is a particular choice of transversal which results in the first return map having as its permutation the *hyperelliptic permutation*

$$\pi(j) = d - j + 1$$

for $j = 1, \ldots, d - 1$ where $d = 2g$. If ω belongs to the hyperelliptic stratum $\mathcal{H}_{\mathrm{hyp}}(g - 1, g - 1)$, then a choice of transversal can result again in the same permutation π with $d = 2g + 1$. Different choices of transversal can result in different permutations.

The Veech example. The Veech example of a minimal nonuniquely ergodic flow we discussed in §4.3.1 can be turned into a nonuniquely ergodic minimal IET on 5 intervals by taking a transversal to the flow. This was done by Keynes-Newton [**104**].

4.5.1. Mixing. In this section, we introduce the notions of *mixing* and *weak mixing* for dynamical systems, and we discuss when they hold for IETs. These are properties stronger than ergodicity, which in some sense express that the system achieves some kind of long-term independence.

Strong mixing. We first introduce the notion of *strong mixing* or just *mixing* for measure-preserving transformations and flows:

Definition 4.5.3. A measure-preserving transformation $T : (X, \mu) \to (X, \mu)$ is *strong mixing* if for any $f_1, f_2 \in L^2(\mu)$

$$(4.5.2) \qquad \lim_{n \to \infty} \int_X (f_1 \circ T^{-n}(x)) f_2(x) d\mu(x) = \int_X f_1 d\mu \int_X f_2 d\mu.$$

Similarly, a measure-preserving flow ϕ_t on a probability measure space (X, μ) is said to be strong mixing if for any $f_1, f_2, \in L^2(X, \mu)$,

$$(4.5.3) \qquad \lim_{t \to \infty} \int_X (f_1 \circ \phi_t(x)) f_2(x) d\mu(x) = \int_X f_1 d\mu \int_X f_2 d\mu.$$

Weak mixing. Strong mixing says that the observations $f_2(x)$ and $f_1(T^n x)$ (or $f_1(\phi_t x)$) become independent of each other as $n \to \infty$ (or $t \to \infty$). Weak mixing is a Cesáro version of this independence.

Definition 4.5.4. A measure-preserving transformation $T : (X, \mu) \to (X, \mu)$ of a probability space is *weak mixing* if for any $f_1, f_2 \in L^2(X, \mu)$, the sequence $\int_X (f_1 \circ T^{-n}(x)) f_2(x) d\mu(x)$ converges in a Cesáro sense to $\int_X f_1 d\mu \int_X f_2 d\mu$; that is,

$$(4.5.4) \qquad \lim_{n \to \infty} \frac{1}{n} \sum_{j=1}^n \left| \int_X f_1 \circ T^{-j}(x) f_2(x) d\mu - \int_X f_1 d\mu \int_X f_2 d\mu \right| = 0.$$

We say a probability measure-preserving flow ϕ_t on (X, μ) is weak mixing if

$$(4.5.5) \qquad \lim_{T \to \infty} \frac{1}{T} \int_0^T \left| \int_X (f_1 \circ \phi_t)(x) f_2(x) d\mu - \int_X f_1 d\mu \int_X f_2 d\mu \right| dt = 0.$$

Exercise 4.17. *Show that strong mixing implies weak mixing which implies ergodicity.*

Rotations of the circle. We start our discussion of the mixing properties of IETs by first showing that rotations of the circles are never weak mixing.

Theorem 4.5.5. *Irrational rotations of the circle are not weak mixing.*

Proof. Let $A = \{e^{2\pi i \theta} : \theta \in [0, 1/4]\}$ and $B = \{e^{2\pi i \theta} : \theta \in [1/2, 3/4]\}$. Let T be an irrational rotation. Since T^{-1} is ergodic with respect to Lebesgue measure, there is $x \in A$ such that

$$(4.5.6) \qquad \lim_{n \to \infty} \frac{1}{n} \sum_{k=1}^n \mathbf{1}_B(T^{-k}(x)) = \frac{1}{4}.$$

4.5. Interval exchange transformations

Since rotations are isometries (4.5.6) implies that the set of $n \in \mathbb{N}$ such that $T^{-n}(A) \cap A = \emptyset$ has upper density at least $\frac{1}{4}$, where we recall (see, for example, Halberstam-Roth [82, page xvii]) that the *upper density* $\overline{d}(S)$ of a subset $S \subset \mathbb{N}$ is defined by

$$\overline{d}(S) = \limsup_{n \to \infty} \frac{\mathrm{card}(S \cap \{1, \ldots, n\})}{n}.$$

This in turn implies that

$$|\mu(T^{-n}(A) \cap A) - \mu(A)^2| = \frac{1}{16}$$

for a set of n of upper density at least $\frac{1}{4}$ which implies T cannot be weak mixing. □

Weak mixing for generic IETs. On the other hand we will see below in Chapter 5 that using renormalization methods, Avila-Forni [19] showed that IETs that do not come from rotations are typically weak mixing.

Lack of strong mixing. We now turn to two theorems due to Katok [97], which prove that interval exchange transformations and translation flows are never strong mixing. Our proof follows that of Katok.

Theorem 4.5.6. *Interval exchange transformations T are not strong mixing.*

Proof. By Lemma 4.5.2 we can assume the invariant measure of T is Lebesgue measure λ. Let $J \subset [0, 1]$ and define T_J to be the first return map of T to J. By Exercise 4.13, T_J is an IET on at most $d+2$ intervals $J_1, \ldots, J_s$, where $J = \bigcup_{i=1}^{s} J_i$, and there are positive integers k_i such that for $i = 1, \ldots, s$,

$$T_J|_{J_i} = T^{k_i}.$$

For any m, set $J_i^m = T^m(J_i)$. Then we can partition the interval $[0, 1)$ as a union of the J_i^m; that is,

$$[0, 1) = \bigcup_{i=1}^{s} \bigcup_{m=0}^{k_i-1} J_i^m.$$

By taking $|J|$ small we make the individual elements of the partition arbitrarily small. Now for each J_i we take the induced interval exchange on J_i. This means we divide J_i into intervals $J_{i,j}$, $j = 1, \ldots, s_i$, and for each $J_{i,j}$ there is a $k_{i,j}$ such that $T^{k_{i,j}}$ gives the first return of $J_{i,j}$ to J_i. This gives a refined partition of $[0, 1)$ by

$$[0, 1) = \bigcup_{i=1}^{s} \bigcup_{j=1}^{s_i} \bigcup_{m=0}^{k_{ij}-1} T^m(J_{ij}).$$

Set $J_{ij}^m = T^m(J_{ij})$ and $J_i^m = T^m(J_i)$. Now for each k_{ij},

$$T^{k_{ij}}(J_{ij}^m) \subset J_i^m$$

so
$$J_{ij}^m \subset T^{-k_{ij}} J_i^m.$$
Further $J_i^m = \bigcup_{j=1}^{s_i} J_{ij}^m$ together giving
$$J_i^m \subset \bigcup_{j=1}^{s_i} T^{-k_{ij}} J_i^m.$$

Now let B be any set which is a union of a collection of sets J_i^m. Then

(4.5.7) $$B \subset \bigcup_{i=1}^{s} \bigcup_{j=1}^{s_i} T^{-k_{ij}}(B).$$

Since T is λ-measure-preserving and $s_i, s_{ij} \leq d+2$, (4.5.7) says there exists k_{ij} such that
$$\lambda(B \cap T^{k_{ij}}(B)) = \lambda(T^{-k_{ij}}(B) \cap B) \geq \frac{1}{(d+2)^2} \lambda(B).$$

Choose a set A such that

(4.5.8) $$\lambda(A) < \frac{1}{10(d+2)^2}.$$

Let N be any positive integer. Choose J small enough so that there is a set B which is a union of partition sets J_i^m such that
$$\lambda(A \Delta B) < \lambda(A)^2/10$$
and the associated numbers $k_i \geq N$. To guarantee that the $k_i \geq N$, we note that the minimality of the transformation T says we may choose J such that
$$T^m(J) \cap J = \emptyset$$
for all $m \leq N$. One gets
$$\lambda(A \cap T^{k_{ij}}(A)) \geq \lambda(B \cap T^{k_{ij}}(B)) - 2\lambda(A \Delta B) \geq \frac{1}{(d+2)^2} \lambda(B) - \frac{1}{5}\lambda(A)^2.$$

Now
$$\lambda(B) \geq \frac{9}{10}\lambda(A)$$
so by (4.5.8)
$$\lambda(A \cap T^{k_{ij}}(A)) \geq \frac{9}{10} \frac{1}{(d+2)^2} \lambda(A) - \frac{1}{5}\lambda(A)^2$$
$$\geq \lambda(A)^2 \left(10 \cdot \frac{9}{10} - \frac{1}{5}\right)$$
$$> 2\lambda(A)^2.$$

Since A was fixed and for any N we have found $k_{ij} \geq N$ such that the above inequality holds, T cannot be strong mixing with respect to λ. □

4.5. Interval exchange transformations

Lack of mixing for flows. Using the relationship between IETs and linear flows, Katok [97] showed:

Theorem 4.5.7. *Let ϕ_t^θ be a linear flow in direction θ on a translation surface ω. Then ϕ_t^θ is not strong mixing.*

In fact, Katok proved a stronger theorem [97] which says that given a flow over an interval exchange, if the return time function is of bounded variation, then the resulting flow is not strong mixing.

Chapter 5

Renormalization

We now turn to one of the crucial themes of this book and, indeed, of the study of translation surfaces:

> *We can obtain results on ergodic properties and counting problems of an individual translation surface by understanding the behavior of its SL(2, ℝ)-orbit in its stratum.*

We start by giving a criterion (§5.1) for unique ergodicity of a vertical flow on a translation surface in terms of the recurrence properties of its orbit under the positive diagonal subgroup A of $SL(2, \mathbb{R})$. We then prove in §5.2 that the flow on a stratum $\mathcal{H}$ defined by the action of this subgroup, known as the *Teichmüller geodesic flow*, is ergodic with respect to the MSV measure $\mu_{\mathcal{H}}$. In the later sections of this chapter, we discuss further developments in this direction, mostly without proofs. We discuss mixing properties of this flow in §5.3, quantitative renormalization results in §5.4, and the size of sets of nonuniquely ergodic linear flows on translation surfaces in §5.5.

5.1. A criterion for unique ergodicity

Notation. We fix notation for the rest of this chapter. We write $\mathcal{H}$ for a connected component of a stratum $\mathcal{H}(\alpha)$ of unit-area translation surfaces, and we let $\mu_{\mathcal{H}}$ denote the MSV measure on $\mathcal{H}$.

Ergodicity and Teichmüller flow. We will now give a sufficient condition for unique ergodicity of the vertical flow on a translation surface ω in terms of its orbit under the positive diagonal subgroup (defined in (3.4.3))

$$A = \left\{ g_t = \begin{pmatrix} e^{t/2} & 0 \\ 0 & e^{-t/2} \end{pmatrix} : t \in \mathbb{R} \right\},$$

and we subsequently use this to show that for $\mu_{\mathcal{H}}$-almost every $\omega \in \mathcal{H}$, the vertical flow is uniquely ergodic. Recall from §3.4.1 that the diagonal action is known as the *Teichmüller geodesic flow*, since projections of orbits to moduli space $\mathcal{M}_g$ are geodesics with respect to the Teichmüller metric.

Masur's criterion. The following theorem of Masur [119] is the first result of this chapter. It shows how *recurrence* properties of the g_t-orbit of ω on $\mathcal{H}$ give information about the ergodic properties of the vertical flow on ω.

Theorem 5.1.1. *Let $\omega \in \mathcal{H}$ and let $\pi : \mathcal{H} \to \mathcal{M}_g$ be the projection to the moduli space of Riemann surfaces. Suppose the Teichmüller geodesic $\{X_t = \pi(g_t\omega) : t > 0\}$ is nondivergent; that is, there is a sequence $t_n \to \infty$ and a compact set $C \subset \mathcal{M}_g$ so that*

$$\pi(g_{t_n}\omega) \in C.$$

Assuming the vertical flow ϕ_t is minimal, then it is uniquely ergodic.

A limiting surface. To prove Theorem 5.1.1, we proceed by contradiction. Suppose that the vertical flow ϕ_t on ω is not uniquely ergodic and that the Teichmüller geodesic $\{X_t = \pi(g_t\omega) : t > 0\}$ is *nondivergent*. By passing to a subsequence, we can take it that as $t_n \to \infty$, the Riemann surfaces $X_n = X_{t_n}$ converge in $\mathcal{M}_g$ to some $X_\infty \in \mathcal{M}_g$. Since the set of unit-area translation surfaces lying over a compact set of $\mathcal{M}_g$ is itself compact, by passing to a further subsequence we may assume that $\omega_n = g_{t_n}(\omega)$ converges to some ω_∞. We can take this to mean that we can choose polygon(s) on X_n defining ω_n that converge to polygons on X_∞ defining ω_∞, in the sense that all the sides converge as complex numbers. Let $f_{t_n} : X \to X_n$ be the corresponding *Teichmüller map* (recall Black Box 3.4.4) from the base surface X of ω. Denote by Z and Z_∞ the sets of zeros of ω and ω_∞, respectively.

Invariant measures. Since we are assuming that ϕ_t is nonuniquely ergodic, we have at least two mutually singular ergodic invariant probability measures ν_i, which are extreme points of the set of invariant probability measures for the flow ϕ_t. Let λ be Lebesgue measure on ω. By Theorem 4.4.4 we have

$$\lambda = \sum_{i=1}^m c_i \nu_i,$$

for a collection of $2 \le m \le g$ ergodic probability measures ν_i and constants $0 < c_i \le 1$. (We note that $m \ge 2$ since we are assuming λ is not uniquely ergodic.) Each measure ν_i can be written locally as

$$\nu_i = \hat{\nu}_i \times dy$$

5.1. A criterion for unique ergodicity

where $\hat{\nu}_i$ is a measure on a horizontal transversal and dy is Lebesgue measure on vertical lines.

Generic points. Let $E_i \subset X$ be the disjoint sets of generic points of the measure ν_i. The set of points that lie on a vertical line that runs into a zero of ω has Lebesgue measure 0, hence ν_i-measure 0 for each i. We remove these points from E_i. By the Birkhoff Ergodic Theorem, Theorem 4.2.4, we have

$$\nu_i(E_i) = 1,$$

and each E_i is invariant under ϕ_t.

Limit sets. For each i, let A_i be the set of all possible accumulation points of $f_{t_n}(x) \in X_n$, where $x \in E_i$ (that is, we take the union over all $x \in E_i$). Now let $U \subset X_\infty$ be any open set. We claim that there is an index j such that

(5.1.1) $$U \cap A_j \neq \emptyset.$$

Proving (5.1.1). To prove (5.1.1), choose an open W so that $\bar{W} \subset U$. Then for large n, since the polygons on X_n converge to those of X_∞, W may be considered as a subset of X_n with area bounded below by $\delta > 0$. Consider

$$W_n = f_{t_n}^{-1}(W) \subset X.$$

Since Teichmüller maps f_{t_n} preserve area (as they come from the action of the diagonal group in $SL(2, \mathbb{R})$),

$$\lambda(W_n) > \delta.$$

Thus for each n, there is a $j = j(n)$ such that

(5.1.2) $$\nu_j(E_j \cap W_n) = \nu_j(W_n) \geq \frac{\delta}{mc_j}.$$

Since by Theorem 4.4.4 there are at most g measures ν_j, we may choose a j so that (5.1.2) holds for infinitely many n. Then there exists some $x \in E_j \cap W_n$ for infinitely many n. Since $f_{t_n}(x) \in W$, the sequence $\{f_{t_n}(x)\}_{n \geq 1}$ has an accumulation point in $\bar{W}$, proving the claim.

Claim 5.1.2. With notation as above, suppose $R_\infty \subset X_\infty$ is a closed rectangle of ω_∞ that does not contain zeros of ω_∞. Suppose $x_\infty \in A_i$ is the lower left vertex and $y_\infty \in A_j$ is the upper right vertex. Then $\nu_i = \nu_j$.

Proving Claim 5.1.2. We prove Claim 5.1.2 by contradiction. By relabeling assume the measures are $\nu_1 \neq \nu_2$. First choose an interval I on the surface X such that

$$\frac{\hat{\nu}_1(I)}{\hat{\nu}_2(I)} \neq 1,$$

where $\hat{\nu}_1, \hat{\nu}_2$ are the transverse ergodic measures on I, and let l_1, l_2 be the vertical sides of R_∞.

Limiting sequences. Let $x_n \in f_{t_n}(E_1)$ limit to x_∞ and let $y_n \in f_{t_n}(E_2)$ limit to y_∞. The points x_n and y_n are vertices of rectangles R_n converging to R_∞; see Figure 5.1. Let $l_{1,n}, l_{2,n} \subset R_n$ be the vertical segments with endpoints x_n and y_n. They limit to l_1 and l_2, respectively. Now $l_{1,n}$ and $l_{2,n}$ have the same Euclidean length. There are points $x \in E_1$ and $y \in E_2$ such that for $i = 1, 2$, $l_{i,n}$ is the image under f_{t_n} of a vertical segment $L_{i,n} \subset E_i$ passing through x and y, respectively. Now $L_{1,n}$ and $L_{2,n}$ have equal Euclidean length on the base surface ω and the lengths go to ∞ with n and so the number of intersections with I goes to ∞. These facts together with the fact that $L_{i,n} \subset E_i$ imply that

$$(5.1.3) \qquad \lim_{n \to \infty} \frac{\#(l_{1,n} \cap f_{t_n}(I))}{\#(l_{2,n} \cap f_{t_n}(I))} = \lim_{n \to \infty} \frac{\#(L_{1,n} \cap I)}{\#(L_{2,n} \cap I)} = \frac{\hat{\nu}_1(I)}{\hat{\nu}_2(I)} \neq 1.$$

Intersections with l_i. Since $l_{i,n}$ has limit $l_i \subset \omega_\infty$ and every horizontal segment of ω_∞ that intersects l_1 intersects l_2 and vice versa and since $f_{t_n}(I)$ is a horizontal line whose length goes to ∞, the ratio

$$\frac{\#(l_{1,n} \cap f_{t_n}(I))}{\#(l_{2,n} \cap f_{t_n}(I))}$$

can be made arbitrarily close to 1 by taking n large enough. This is a contradiction to (5.1.3), concluding the proof of Claim 5.1.2. The proof of the theorem will follow from Claim 5.1.2 and the following global argument.

Global argument. We now follow an argument of Monteil [132] that replaces the original global argument in [119] (see also the exposition of the original proof in [120]). We assumed above that the limit points x_∞, y_∞ lie in a rectangle that does not include any zeros. Suppose first that either one, say x_∞, is a zero of ω_∞. It is the limit of $x_n = f_{t_n}(x)$ where x is generic for ν_1. The point x does not lie on a vertical line that ends at a zero so replacing the flow time t_n with a small variation t'_n, we can assume $f_{t'_n}(x)$ stays a uniformly bounded distance away from the zeros of ω_n and so we can assume x_∞ and y_∞ are not zeros of ω_∞.

Chains of rectangles. Next take a path σ from x_∞ to y_∞ and an open set U containing σ that does not contain any zeros of ω_∞. The compactness of the surface implies that there is a finite sequence of successive points $x_\infty = x_\infty^1, x_\infty^2, \ldots, x_\infty^k = y_\infty$ along σ such that each successive pair x_∞^j and x_∞^{j+1} are vertices of a rectangle contained in U. By applying (5.1.1), by moving the point x_∞^j, if necessary, a small distance, we can assume x_∞^j is still contained in the open set U and lies in one of the sets A_i for some measure ν_i. It will still be the vertex of rectangle contained in U. Then applying the main argument we have that for each successive pair of points the measures coincide and so they coincide for x_∞ and y_∞. Since these points were arbitrary points in $\bigcup A_i$ there is a single value of i such that all points lie in A_i, which gives a contradiction to the assumption that ϕ_t is not uniquely ergodic.

5.1. A criterion for unique ergodicity

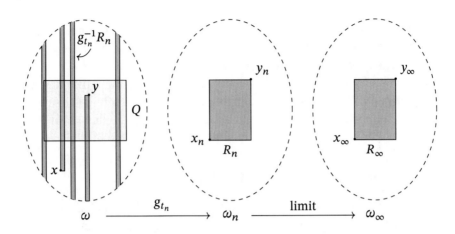

Figure 5.1. The limit surface ω_∞ and the limiting procedure, renormalizing a long thin rectangle. Tikz code for this figure courtesy of T. Monteil [**132**], with permission from Mathematisches Forschungsinstitut Oberwolfach.

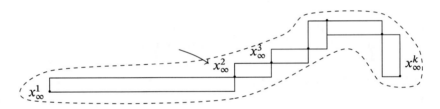

Figure 5.2. Chains of rectangles. Tikz code for this figure courtesy of T. Monteil [**132**], with permission from Mathematisches Forschungsinstitut Oberwolfach.

5.1.1. Almost sure unique ergodicity.
We can combine Theorem 5.1.1 with the classical Poincaré recurrence theorem to prove that for generic ω, the vertical flow on ω is uniquely ergodic. That is, we have

Theorem 5.1.3 ([**119**]). *For $\mu_{\mathcal{H}}$-almost every $\omega \in \mathcal{H}$ the vertical flow ϕ_t is uniquely ergodic.*

Poincaré recurrence. We first recall the Poincaré recurrence theorem for measure-preserving flows, whose proof we leave as Exercise 5.1.

Theorem 5.1.4. *Let ψ_t denote a measure-preserving flow on a finite measure space (X, μ). Then for any $E \subset X$ with $\mu(E) > 0$,*

$$(5.1.4) \qquad \mu(x \in E : \exists T > 0 \text{ such that } \psi_t(x) \notin E \text{ for all } t > T) = 0.$$

Exercise 5.1. *Prove the Poincaré recurrence theorem, Theorem 5.1.4.*

Applying Poincaré recurrence. We recall that Theorem 3.7.1 states that

$$\mu_{\mathcal{H}}(\mathcal{H}) < \infty.$$

Since the Teichmüller geodesic flow g_t is $\mu_{\mathcal{H}}$-measure-preserving, we can apply (5.1.4) to an exhaustion of $\mathcal{H}$ by positive measure compact sets $C_n \subset \mathcal{H}$. Then $\mu_{\mathcal{H}}$-almost every $\omega \in \mathcal{H}$ belongs to some C_n and recurs to C_n under the flow g_t for arbitrarily large times. By Theorem 5.1.1, we know that any ω with nonuniquely ergodic vertical foliations has divergent g_t-orbit, so the set of such ω must be a 0-measure subset of each C_n, hence a 0-measure subset of $\mathcal{H}$ proving Theorem 5.1.3. □

5.2. Ergodicity of the Teichmüller flow

The fact that the g_t-action is $\mu_{\mathcal{H}}$-preserving leads to the natural question of the ergodicity of the action. Theorem 5.2.1 below, independently proven by Masur [118] and Veech [162], shows this ergodicity.

Theorem 5.2.1. *The action of A (the Teichmüller geodesic flow) on $\mathcal{H}$ is ergodic with respect to MSV measure $\mu_{\mathcal{H}}$.*

The Hopf argument. The rest of this section is devoted to the proof of Theorem 5.2.1. We follow the strategy of Hopf [86] in his proof of the ergodicity of the geodesic flow on the unit tangent bundle of a hyperbolic surface. For a modern treatment of this argument, see, for example, Coudène [39, §4]. Informally, the Hopf argument shows that ergodic averages along g_t-orbits are constant along *stable manifolds* (these are manifolds of points which have asymptotically the same future under g_t), and then it uses this to conclude that g_t-invariant functions must be constant along stable manifolds. To implement this strategy, we first let d_T be the Teichmüller metric and $d_{\mathcal{H}}$ be a metric on the stratum compatible with the topology arising from period coordinates.

Measured foliations. We refer to Farb-Margalit [64, §11] as a basic reference here. To define the stable manifolds in our setting, we first recall the notion of a *measured foliation* of a surface S. Our discussion follows [64, §11.2]. A *measured foliation* is a singular foliation $\mathcal{F}$ with a nontrivial transverse-invariant measure ν. A *singular foliation* $\mathcal{F}$ on a closed surface S decomposes S as a disjoint union of 1-dimensional *leaves* and a finite set of *singular points*. If $p \in S$ is nonsingular, there is a smooth chart $\phi : U \to \mathbb{R}^2$ from a neighborhood of U containing p that takes leaves of $\mathcal{F}$ to horizontal line segments, and the transition maps between these charts preserve the horizontal direction. At singular points $q \in S$, there is a smooth chart $\phi : V \to \mathbb{R}^2$ from a neighborhood V of q to $\mathbb{R}^2$ that takes leaves of $\mathcal{F}$ to the level sets of a saddle with at least 3 prongs. The measure ν assigns a nonnegative real number for each transverse arc in a countably additive fashion, and that assigns 0 to an arc if and only if the arc lies in a leaf. Transverse invariance means that the measure ν does not change if a transverse arc is moved to another transverse arc with endpoints in the same leaves of $\mathcal{F}$. If the measured foliation is *orientable*, then it defines a flow.

5.2. Ergodicity of the Teichmüller flow

Whitehead equivalence. Two measured foliations are called *Whitehead equivalent* if they are related by a finite sequence of isotopies that preserve the transverse measure and by *Whitehead moves*. In a Whitehead move a measured foliation has a compact leaf joining two singular points and the leaf is collapsed to a point, while preserving the transverse measure on the complement. If the singular points denoted p and q being collapsed to a single point w have p and q prongs, respectively, then q has $p + q - 2$ prongs.

The vertical flow. To each translation surface ω, we have a natural oriented measured foliation associated to any invariant measure η for the vertical flow. Namely, the flow is given by $\{\text{Re}(\omega) = 0\}$, and the transverse-invariant measure $\nu = \hat{\eta}$ associated to η is given by distintegrating η as in §5.1. If the vertical flow is uniquely ergodic, there is a unique transverse-invariant measure and thus a unique measured foliation associated to the flow. We will say that two translation surfaces ω_1, ω_2 with uniquely ergodic vertical flows have the same vertical flow if the associated measured foliations are Whitehead equivalent. We can also define the *horizontal foliation* associated to ω by considering the leaves $\{\text{Im}(\omega) = 0\}$. The following theorem of Masur [117] says that orbits with the same vertical uniquely ergodic flow are asymptotic.

Theorem 5.2.2. *Suppose $\omega_1, \omega_2 \in \mathcal{H}$ have the same uniquely ergodic vertical flow. Then*
$$\lim_{t \to \infty} d_T(\pi(g_t \omega_1), \pi(g_t \omega_2)) = 0.$$
Here, π denotes the projection from the stratum to Teichmüller space.

We remark that the same theorem holds in the more general context of quadratic differentials where one may have a general measured foliation and not a flow.

Zippered rectangles. In the case of a flow which we are considering, the first return map to a horizontal interval induces an interval exchange transformation, a concept already discussed. In the case of foliations arising from quadratic differentials, there is a similar concept of a generalized IET, a notion not discussed in this book.

Since the vertical flows are the same and the transverse measures are the same we can choose a common horizontal transversal J for the vertical flows on ω_1, ω_2, and we can assume one endpoint $p \in J$ is a zero. We can assume J has unit length. The first return of the vertical flow gives interval exchange transformations T_1, T_2 defined on J on ℓ intervals J_i. Since the transverse measures are the same, $T_1 = T_2$, so we get zippered rectangle decompositions for ω_1, ω_2 where the horizontal sides are the same but the vertical sides differ in length. Each subinterval $J_i \subset J$ is a horizontal side of a rectangle with heights $H_i(\omega_1)$ and $H_i(\omega_2)$, respectively, for the two translation surfaces. The H_i are the return times for a flow line leaving a point of J_i to return to J.

Inducing on a subinterval. Now consider a subinterval $J' \subset J$ with one endpoint the same zero p and which defines subintervals J'_i of the corresponding IET, denoted T'. This gives new zippered rectangle decompositions of ω_1 and ω_2. Let $H'_i(\omega_1), H'_i(\omega_2)$ be the corresponding heights of the new rectangles. Given a new subinterval J'_i, let $v_{i,k}$ be the number of visits of points of J'_i under the first return map $T_1 = T_2$ to J_k before they return to J'. Then set

$$v_i = \sum_k v_{i,k}$$

to be the total number of visits. The unique ergodicity of the vertical flow implies that as orbits get longer and longer they become equidistributed with the proportion visiting J_k converging to the proportion of $|J_k|$ to $|J| = 1$. Namely

(5.2.1) $$\lim_{|J'|\to 0} \frac{v_{i,k}}{v_i} = |J_k|.$$

Area 1. From this we conclude that for both translation surfaces ω_1 and ω_2

(5.2.2) $$\lim_{|J'|\to 0} \sum_k H_k(\omega_i)\left(\frac{v_{i,k}}{v_i}\right) = \sum_k H_k|J_k| = 1,$$

the area of the surfaces ω_i.

Computing new heights. The new heights H'_i can be expressed in terms of the original heights and the visits. We get that for each i,

(5.2.3) $$H'_i(\omega_1) = \sum_k v_{i,k} H_k(\omega_1) \quad \text{and} \quad H'_i(\omega_2) = \sum_k v_{i,k} H_k(\omega_2).$$

Together with (5.2.2), (5.2.3) implies

$$\lim_{|J'|\to 0} \frac{H'_i(\omega_1)}{H'_i(\omega_2)} = 1.$$

Widths and ratios. This says that if we choose J' to be a small enough transversal, the ratio of heights of the rectangles in the new zippered rectangular decomposition can be made as close to 1 as desired, while the widths remain equal. We now argue that the same argument holds for the ratio of the distance from the vertices of the rectangles which lie on J'_i to the zeros on the vertical sides. Namely for each J'_i we can count the number of visits to intervals J_n until a vertical side of J'_i runs into a zero on the vertical side of some J_k. The number of visits goes to infinity as the length of J' goes to 0 and since the first return maps for ω_1 and ω_2 are the same the number of visits coincide. As in the previous argument this gives the desired statement about lengths.

Applying Teichmüller flow. Now apply Teichmüller geodesic flow g_t for large t. The horizontal sides of the rectangles remain equal for ω_1 and ω_2 and they become long while the vertical sides become short, although for sides of corresponding rectangles for $g_t(\omega_1)$ and $g_t(\omega_2)$ with ratio approaching 1 as $t \to \infty$, see Figure 5.3.

5.2. Ergodicity of the Teichmüller flow

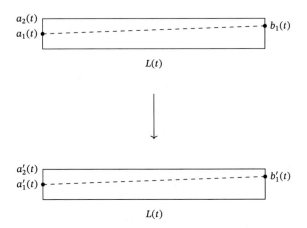

Figure 5.3. Illustration of the rectangle map.

Building a map of rectangles. We show there is a map $F(t, x, y)$ from each rectangle of $g_t\omega_1$ to the corresponding rectangle of $g_t\omega_2$ which takes vertices to vertices and takes zeros on the sides to zeros on the sides such that the dilatation $K(F)$ is arbitrarily close to 1. These rectangle maps then can be glued together to build an almost 1-quasiconformal map of one zippered rectangle decomposition to the other, hence a map between corresponding Riemann surfaces $X_t = \pi(g_t\omega_1)$ and $Y_t = \pi(g_t\omega_2)$.

Ratios. We refer to Figure 5.3. The vertices of the rectangles lie on the transversal J'_i. The black dots correspond to zeros. The heights of the rectangles are $a_2(t), a'_2(t)$ and the distances of the zeros on the sides to the lower left and lower right vertices are $a_1(t), a'_1(t), b_1(t), b'_1(t)$. Since the ratios of the heights of rectangles and distances of zeros to vertices go to 1, we have that

$$\lim_{t\to\infty} \frac{a_1(t)}{a'_1(t)} = \lim_{t\to\infty} \frac{a_2(t)}{a'_2(t)} = \lim_{t\to\infty} \frac{b_1(t)}{b'_1(t)}$$
$$= \lim_{t\to\infty} \frac{a_2(t) - a_1(t)}{a'_2(t) - a'_1(t)}$$
$$= \lim_{t\to\infty} \frac{a_2(t) - b_1(t)}{a'_2(t) - b'_1(t)} = 1$$

and

$$\lim_{t\to\infty} a_2(t) = 0.$$

In addition, we have

$$\lim_{t\to\infty} L(t) = \infty.$$

Trapezoids. Consider the trapezoids Ω, Ω' below the dotted line in the rectangles in Figure 5.3. We define $F(t, x, y) = (u(x, y), v(x, y)) : \Omega \to \Omega'$ by

$$u(t, x, y) = x, \quad v(t, x, y) = \frac{a_1'(t) + \frac{b_1'(t) - a_1'(t)}{L(t)} x}{a_1(t) + \frac{b_1(t) - a_1(t)}{L(t)} x} y.$$

One checks that in fact $F(t, x, y)$ is a homeomorphism. It maps the dotted lines to each other, preserving the x-coordinate. A computation using the assumptions on ratios of lengths shows

$$\partial u(t, x, y)/\partial x = 1,$$
$$\lim_{t \to \infty} \partial v(t, x, y)/\partial y = 1,$$
$$\lim_{t \to \infty} \partial v(t, x, y)/\partial x = 0,$$
$$\partial u(t, x, y)/\partial y = 0,$$

which implies that the quasiconformal dilatation

$$\lim_{t \to \infty} K(F(t, x, y)) = 1.$$

We similarly have a map $G(t, x, y)$ of the trapezoids above the dotted lines to each other with $K(G(t, x, y)) \to 1$. The maps glue together along the dotted line to give a map of each rectangle and then the maps are glued along rectangles to give a map $X_t \to Y_t$ with dilatation approaching 1. By the definition of the Teichmüller metric this finishes the proof.

Invariance of ergodic averages. We continue the proof of Theorem 5.2.1. For $\omega \in \mathcal{H}, f \in L^1(\mathcal{H}, \mu)$, define

(5.2.4) $$f^*(\omega) = \lim_{T \to \infty} \frac{1}{T} \int_0^T f(g_t \omega) dt$$

and

(5.2.5) $$f_*(\omega) = \lim_{T \to \infty} \frac{1}{T} \int_0^T f(g_{-t} \omega) dt$$

to be the forward and backward ergodic averages of f along the g_t-orbit of ω. It is easy to see that if h is a bounded g_t-invariant function, then $(fh)^* = f^* h$ and $(fh)_* = f_* h$. Since $\mu_{\mathcal{H}}(\mathcal{H}) < \infty$, the Birkhoff Ergodic Theorem 4.2.4 says $(fh)^*(\omega)$ and $(fh)_*(\omega)$ exist for $\mu_{\mathcal{H}}$-almost every ω and

(5.2.6) $$\int_{\mathcal{H}} f^* h d\mu = \int_{\mathcal{H}} (fh)^* d\mu = \int_{\mathcal{H}} (fh)_* d\mu = \int_{\mathcal{H}} f_* h d\mu.$$

Taking $h = \text{sgn}(f^* - f_*)$ we have $f^* = f_*$ almost everywhere. To conclude that the flow is ergodic, we need to show that f^* (and thus f_*) are almost everywhere constant.

5.2. Ergodicity of the Teichmüller flow

Asymptotic orbits. We first prove that f^* is constant along what are called stable manifolds. Assume $\omega \in \mathcal{H}$ and suppose the vertical flow of ω is uniquely ergodic. Let $\text{Stab}(\omega)$ be the set of $\hat{\omega} \in \mathcal{H}$ such that for some constant $\lambda > 0$, $\text{Re}(\hat{\omega}) = \lambda \text{Re}(\omega)$; that is, $\hat{\omega}$ has the same vertical foliation as ω.

Lemma 5.2.3. *Let f be continuous on $\mathcal{H}$ with compact support $C \subset \mathcal{H}$. Suppose $\omega_1, \omega_2 \in \mathcal{H}$ and $\omega_2 \in \text{Stab}(\omega_1)$. Suppose $f^*(\omega_1)$ exists. Then so does $f^*(\omega_2)$ and $f^*(\omega_1) = f^*(\omega_2)$. The analogous statement holds for a pair of surfaces with the same uniquely ergodic horizontal flow and f_*.*

It is a consequence of [88, Main Theorem] that the map $\pi : \text{Stab}(\omega) \to T_g$ is a homeomorphism onto its image, where, as in the statement of Theorem 5.2.2, we are using π to denote the projection from a stratum to Teichmüller space. Since f is continuous with compact support, for any ϵ there is a δ such that if $\hat{\omega} \in \text{Stab}(\omega)$ and $d_T(\pi(\hat{\omega}), \pi(\omega)) < \delta$, then

$$(5.2.7) \qquad |f(\hat{\omega}) - f(\omega)| < \epsilon.$$

Assume first that ω_2 is in the strong stable manifold of ω_1. This means that $\lambda = 1$, or equivalently ω_1 and ω_2 have the same transverse measure. By Theorem 5.2.2, for all $\delta > 0$ there exists T_0 such that for $t > T_0$

$$d_T(\pi(g_t\omega_1), \pi(g_t\omega_2)) < \delta.$$

Now letting $\omega = g_t\omega_1$ and $\hat{\omega} = g_t\omega_2$ and applying (5.2.7) we see

$$|f(g_t\omega_1) - f(g_t\omega_2)| < \epsilon$$

for $t > T_0$. Let $B = \max_{\omega \in C} |f(\omega)|$ be the maximum of f on C, and define the error function E by

$$E(t) = |f(g_t\omega_1) - f(g_t\omega_2)|,$$

$$\left| \frac{1}{T} \int_0^T (f(g_t\omega_1) - f(g_t\omega_2)) dt \right| \leq \frac{1}{T} \left(\int_0^{T_0} E(t) dt + \int_{T_0}^T E(t) dt \right)$$

$$\leq \frac{2BT_0}{T} + \frac{\epsilon(T - T_0)}{T}.$$

Letting $T \to \infty$ and noting that ϵ was arbitrary shows $f^*(\omega_1) = f^*(\omega_2)$. Now since f^* and f_* are g_t-invariant, the lemma holds for $\omega_2 \in \text{Stab}(\omega_1)$.

Linear operators. We now note following Hopf [86] that the linear operator

$$f \mapsto f^*$$

is a bounded operator on $L^1(\mathcal{H}, \mu_{\mathcal{H}})$. This follows from (5.2.6) by taking $h = \text{sgn } f^*$, giving

$$\int_{\mathcal{H}} |f^*| d\mu_{\mathcal{H}} \leq \int_{\mathcal{H}} |f| d\mu_{\mathcal{H}}.$$

Therefore to finish the proof of Theorem 5.2.1 it is enough to show that f^* is constant almost everywhere for a dense set of $f \in L^1$, namely those f continuous with compact support. To prove this, we have the following lemma.

Lemma 5.2.4. *Suppose $f : \mathcal{H} \to \mathbb{R}$ is continuous with compact support. For any $k > 0$ let*
$$M_f = \{\omega : f^*(\omega) \geq k\}$$
and
$$M_b = \{\omega : f_*(\omega) \geq k\}.$$
Then either $\mu_\mathcal{H}(M_f) = \mu_\mathcal{H}(M_b) = 0$ or $\mu_\mathcal{H}(\mathcal{H} \setminus M_f) = \mu_\mathcal{H}(\mathcal{H} \setminus M_b) = 0$.

Proof. Recall that we can describe translation surfaces locally by period coordinates $\left(\int_{\gamma_i} \omega\right)_{i=1}^n$ with respect to a basis $\{\gamma_i : i = 1, \ldots, n\}$ for the relative homology group $H_1(S, \Sigma, \mathbb{Z})$. Breaking the coordinates into real and imaginary parts we can assume that ω_1, ω_2 lie in a rectangle $J_1 \times J_2 \subset \mathbb{R}^n \times \mathbb{R}^n$ where J_1 records the real part of the periods and J_2 the imaginary part. Since g_t expands one transverse measure and contracts the other, we can describe the set of orbits of ϕ_t in $J_1 \times J_2$ as a product of projective spaces $\mathbb{P}(J_1) \times \mathbb{P}(J_2)$. We can therefore write
$$M_f = E_1 \times \mathbb{P}(J_2)$$
for some $E_1 \subset \mathbb{P}(J_1)$, and similarly $M_b = \mathbb{P}(J_1) \times E_2$ for $E_2 \subset \mathbb{P}(J_2)$. Since $f^* = f_*$ almost everywhere and each is invariant under the flow, we have
$$m(M_f \setminus M_b) = 0 \quad \text{and} \quad m(M_b \setminus M_f) = 0,$$
where m is measure class on $\mathbb{P}(J_1) \times \mathbb{P}(J_2)$. This gives
$$m(E_1 \times E_2^c) = 0 = m(E_1^c \times E_2),$$
and from that either $m(E_1^c) = m(E_2^c) = 0$ or $m(E_1) = m(E_2) = 0$ where m is measure class on $\mathbb{P}(J_i)$. $\square$

5.3. Geodesics, horocycles, and mixing

Theorem 5.2.1 has several important implications for the full $SL(2, \mathbb{R})$-action. In particular, we have

Theorem 5.3.1. *The $SL(2, \mathbb{R})$-action on $\mathcal{H}$ is ergodic with respect to the measure $\mu_\mathcal{H}$.*

Proof. Any $SL(2, \mathbb{R})$-invariant function f on $\mathcal{H}$ is g_t-invariant, and so by Theorem 5.2.1 it is $\mu_\mathcal{H}$-almost everywhere constant. $\square$

5.3. Geodesics, horocycles, and mixing

5.3.1. The Howe-Moore theorem. The advantage of having a large group like $SL(2, \mathbb{R})$ acting ergodically is that this ergodicity (and indeed, mixing) passes down to noncompact subgroups, due to a remarkable theorem of Howe-Moore [87]. We record a version of this theorem as a Black Box. For an excellent exposition of the Howe-Moore theorem and its applications, see Bekka-Mayer [25] or Einsiedler-Ward [52].

Black Box 5.3.2. *Let G be a semisimple Lie group acting ergodically on a probability measure space (X, μ). Then for any noncompact subgroup H, the induced action of H on (X, μ) is ergodic, and in fact it is mixing: for any mean 0 functions $\varphi, \psi \in L_0^2(X, \mu)$ and any sequence $h_n \in H$ with $h_n \to \infty$*

$$(5.3.1) \qquad \lim_{n \to \infty} \int_X \varphi(h_n x) \psi(x) d\mu(x) = 0.$$

The horocycle flow. An immediate consequence of Black Box 5.3.2 is the ergodicity of the *horocycle flow* on strata of translation surfaces, given by the action of the subgroup

$$U = \left\{ h_s = \begin{pmatrix} 1 & s \\ 0 & 1 \end{pmatrix} : s \in \mathbb{R} \right\}.$$

5.3.2. Orbit closures. Following the work of Marina Ratner [140], who studied orbit closures and invariant measures for the actions of Lie groups on homogeneous spaces (see Morris's book [133] for a careful exposition), we can ask about classifying orbit closures and invariant measure for the action of $SL(2, \mathbb{R})$ and its subgroups on $\mathcal{H}$.

Hyperbolic surfaces. For context, we first discuss an important motivating example of an $SL(2, \mathbb{R})$-action. Dal'Bo's introductory book [40] or Einsiedler-Ward [52] are excellent sources for detailed discussions of this material. The group $SL(2, \mathbb{R})$ acts *transitively* on the unit tangent bundle of a hyperbolic surface, and if the surface has finite volume, it preserves the finite Haar measure. Thus both the geodesic flow (the action of A) and the horocycle flow (the action of U) are *ergodic* (these actions give their names to the flows, as the orbits are geodesics and horocycles, respectively).

The geodesic flow. The geodesic flow is quite unruly, as a famous folk theorem of Furstenberg-Weiss asserts that for any real $1 \leq \alpha \leq 3$ there is a geodesic flow-invariant measure whose support has Hausdorff dimension α.

Horocycle flow. In contrast, the horocycle flow is more rigid: if the hyperbolic surface is compact, Furstenberg [75] showed that the horocycle flow is *uniquely ergodic*. If the surface is noncompact, Dani-Smillie [41] showed that every horocycle orbit is either closed or equidistributed.

Horocycle flow orbit closures on strata. While the orbits of horocycles on hyperbolic surfaces are well-behaved, the orbits of the unipotent flow on strata of translation surfaces can be quite complicated; for example, we have the following result of Chaika-Smillie-Weiss [33].

Theorem 5.3.3 ([33]). *There are horocycle orbits on the stratum $\mathcal{H}(2)$ with orbit closure with fractional Hausdorff dimension.*

Nondivergence. While the closures of horocycles can display unusual behavior, orbits still have strong *recurrence* properties, as studied by Minsky-Weiss [129]. An important special case of their results is

Theorem 5.3.4 ([129]). *There are no divergent orbits for the Teichmüller horocycle flow.*

$SL(2,\mathbb{R})$**-orbit closures.** When we consider the full group $SL(2,\mathbb{R})$ (or even the subgroup $P = AU$ of upper-triangular matrices, orbit closures and invariant measures once again become much more regular, due to pioneering work of Eskin-Mirzakhani [61] and Eskin-Mirzakhani-Mohammadi [62], building on work of McMullen [125] in the genus 2 setting. While their groundbreaking proofs are far beyond the scope of our book, we record these results as a Black Box:

Black Box 5.3.5. *Any P-invariant probability measure on $\mathcal{H}$ is in fact $SL(2,\mathbb{R})$-invariant. Moreover, this measure is* affine: *it is supported on an immersed submanifold which is locally defined by linear equations in period coordinates and is, in an appropriate sense, the pullback of Lebesgue measure in these coordinates. Moreover, any $SL(2,\mathbb{R})$-orbit closure is an affine invariant submanifold and supports such an invariant measure.*

Applications and further reading. We will discuss in Chapter 6 many applications of the ergodic theory of $SL(2,\mathbb{R})$ to counting problems on translation surfaces and, in particular, how the results of [62] can give asymptotics for counting saddle connections (Theorem 6.6.1) for every translation surface. For further reading on orbit closures and related results, the interested reader is urged to consult the surveys of Zorich [178] and Wright [173].

5.3.3. Exponential mixing. The Howe-Moore theorem (Black Box 5.3.2) implies that both the Teichmüller geodesic and horocycle flows are mixing, but it does not provide *quantitative* information about the convergence rate in (5.3.1), which is often important in applications to rates of convergence in equidistribution theorems and subsequent applications to counting (see §6.4 for example). An important result of Avila-Gouëzel-Yoccoz [22] is that the Teichmüller geodesic flow is *exponentially mixing*. Precisely, we have

Theorem 5.3.6 ([22, Main Theorem]). *The Teichmüller flow on $\mathcal{H}$ is exponentially mixing for Hölder observables. That is, there is a $\delta > 0$ such that under appropriate conditions on the mean 0 test functions φ, ψ,*

$$(5.3.2) \qquad \int_{\mathcal{H}} \varphi(g_t \omega) \psi(\omega) d\mu_{\mathcal{H}}(\omega) = O(e^{-\delta t}),$$

where the implied constant depends on a certain norm (which combines Hölder and L^p information) of the functions φ, ψ.

Representation theory and spectral gaps. We conclude this section by mentioning that there is a deep connection between the rate of mixing of the g_t-action and the representation theory of $SL(2, \mathbb{R})$. To each invariant probability measure ν for the $SL(2, \mathbb{R})$-action on $\mathcal{H}$, we have a *unitary representation* of $SL(2, \mathbb{R})$ on $L^2(\mathcal{H}, \nu)$, that is, a homomorphism

$$\rho : SL(2, \mathbb{R}) \to U(L^2(\mathcal{H}, \nu))$$

from $SL(2, \mathbb{R})$ to the group $U(L^2(\mathcal{H}, \nu))$ of unitary operators on $L^2(\mathcal{H}, \nu)$, given by

$$\rho(g)f(\omega) = f(g^{-1}\omega).$$

Any unitary representation of $SL(2, \mathbb{R})$ can be decomposed into a direct integral of irreducible representations (see, for example, Lang [111, §1.2]), and the exponential mixing of the g_t-action is *equivalent* to the absence of a neighborhood of the trivial representation occurring in this direct integral, a phenomenon known as *spectral gap*. This is linked to the spectrum of the associated Laplacian defined by the Casimir operator on $SL(2, \mathbb{R})$. Avila-Gouëzel [21] proved that in fact any ergodic $SL(2, \mathbb{R})$-invariant probability measure has a spectral gap, and so the geodesic flow is mixing with respect to any such invariant measure.

5.4. Quantitative renormalization and ergodicity

Quantitative renormalization. We have seen via Theorem 5.1.1 that *qualitative* behavior of orbits under the Teichmüller geodesic flow (nondivergence) can yield qualitative information about linear flows on translation surfaces (unique ergodicity). In this section we describe (without proofs) further convergence properties for linear flows on translation surfaces, describing some of the implications of fundamental work of Zorich [176], Kontsevich-Zorich [108, 109], Forni [69], and Avila-Viana [23], which makes this renormalization dictionary *quantitative*. This work centers around what has become known as the *Kontsevich-Zorich cocycle* [108, 109], a *symplectic cocycle* over the $SL(2, \mathbb{R})$-action on strata, and its Lyapunov exponents, numbers that measure its rate of exponential growth. This cocycle is a fundamental object in Teichmüller dynamics, and while detailed definitions and proofs are mostly beyond the scope of our book, we will, in our discussions, try to provide precise references as far as is possible. We strongly encourage the interested reader to consult the surveys of Forni [70], Forni-Matheus [71], Viana [166], and Zorich [177].

Cycles and transversals. We start by recalling some ideas from Chapter 4. Given a translation surface ω so that the vertical flow ϕ_t is uniquely ergodic (and so minimal), choose a transversal I to the flow. Given a point $x_0 \in I$ using return points $x_n = \phi_{t_n} x_0 \in I$ to I to form curves γ_n as in §4.4.1, we construct a sequence $\gamma_n \in H_1(X, \mathbb{R})$ converging to the Schwartzmann asymptotic cycle $\lambda \in H_1(X, \mathbb{R})$. This limit does depend on the choice of the transversal (though not on the initial point x_0).

Exercise 5.2. *Show (using unique ergodicity) that if we replace the transversal I with a different transversal J we obtain a colinear asymptotic cycle $\lambda' = c\lambda, c \in \mathbb{R}$.*

Typical translation surfaces. Exercise 5.2 shows that the line L_1 determined by $\lambda \in H_1(X, \mathbb{R})$ is well-defined independent of the choice of transversal. The direction of γ_n converges to the direction of L_1, but in fact, this convergence happens in a very structured way for *typical* flat surfaces ω, in particular, for almost every surface in every connected component $\mathcal{H}$ of a stratum.

5.4.1. Zorich phenomenon.

Numerical experiments. The story of understanding this convergence behavior and its connection to renormalization started with numerical experiments of Zorich [175, 176] and Kontsevich-Zorich [108, 109] on the *deviation of ergodic averages* for interval exchange maps. They explored the number of visits

$$N_T(I, N) = \#\{0 \leq n \leq N : T^n x \in I\}$$

for an orbit $\{T^n x : 0 \leq n \leq N\}$ of a typical (uniquely ergodic) IET T of four or more intervals to a subinterval $I \subset [0, 1]$. Subtracting the expected number of visits $N|I|$, Zorich [175] observed that there appeared to be power-law behavior, namely, that there was a $1 > \nu_2 > 0$ (depending only on the underlying permutation of T) such that

$$\lim_{N \to \infty} \frac{\log |N_T(I, N) - N|I||}{\log N} = \nu_2.$$

Deviation for rotations and 3-IETs. For IETs of two or three intervals, which are rotations or first return maps of rotations, respectively, it follows from the classical Denjoy-Koksma inequality (see Herman [84, page 73]) that this $\nu = 0$.

5.4. Quantitative renormalization and ergodicity

Asymptotic flags. The experiments in fact suggested a more comprehensive picture of further *quantitative information* for typical translation surfaces. Consider the projection Proj_1 of $H_1(X, \mathbb{R})$ to the hyperplane $L_1^\perp$ symplectically orthogonal to L_1. In the case of the torus, $H_1(X, \mathbb{R})$ is 2-dimensional so this hyperplane is a line, and it follows (as mentioned above) from the classical Denjoy-Koksma inequality mentioned above that the length of the projection of γ_n to this line stays bounded. In general $H_1(X, \mathbb{R})$ is 2g-dimensional so the orthogonal hyperplane $L_1^\perp$ is $(2g - 1)$-dimensional and so the projection of γ_n has more flexibility and may be unbounded. The numerical experiments on IETs suggested that for $\mu_\mathcal{H}$-almost every ω there is a direction w_2 in the hyperplane $L_1^\perp$ so that the direction of $\text{Proj}_1(\gamma_n)$ converges to w_2 at a polynomial rate, that is, that the vectors γ_n deviate from the direction λ not arbitrarily, but in a fixed direction w_2, and that, moreover, there is a constant $\nu_2 < 1$ so that

$$\limsup_{n \to \infty} \frac{\log \| \text{Proj}_1(\gamma_n) \|}{\log n} = \nu_2 < 1.$$

Further deviations. The experiments showed that the picture kept repeating at different scales (leading to what Kontsevich-Zorich [108,109] dubbed a *fractal Hodge structure*): letting $L_2 \subset H_1(X, \mathbb{R}) \cong \mathbb{R}^{2g}$ be the 2-dimensional subspace spanned by L_1 and w_2 and letting Proj_2 be the projection to the codimension 2 subspace $L_2^\perp$ symplectically orthogonal to L_2, numerical experiments further suggested that $\text{Proj}_2(\gamma_n)$ converges to a direction w_3 in this subspace and there is a $0 < \nu_3 < \nu_2$ so that

$$\limsup_{n \to \infty} \frac{\log \| \text{Proj}_2(\gamma_n) \|}{\log n} = \nu_3 < \nu_2.$$

This process can be continued for g steps, yielding what is known as an *asymptotic flag*. This led to the following theorem of Zorich [176, Theorem 2], which proved the existence of this flag and (nonstrict) inequalities between the constants ν_i. Fix $\mathcal{H}$ to be a stratum of area 1 genus g translation surfaces, and fix $\mu = \mu_\mathcal{H}$ to be MSV measure on $\mathcal{H}$.

Theorem 5.4.1 ([176, Theorem 2]). *There are numbers $1 > \nu_2 \geq \nu_3 \geq \cdots \geq \nu_g$ ($\nu_i = \nu_i(\mu_\mathcal{H})$) such that for $\mu_\mathcal{H}$-almost every $\omega \in \mathcal{H}$ there exists a flag of subspaces $L_1 \subset L_2 \subseteq \cdots \subseteq L_g \subseteq H_1(X, \mathbb{R})$ in the first homology group $H_1(X, \mathbb{R})$ of the surface with the following properties. Choose any starting point x_0 in a horizontal segment I. The sequence $\gamma_1, \gamma_2, \ldots,$ of cycles satisfies*

$$\lim_{n \to \infty} \frac{\gamma_n}{t_n} = \lambda$$

where t_n is the normalizing sequence defined in (4.4.4), and the nonzero Schwartzmann asymptotic cycle $\lambda \in H_1(X, \mathbb{R})$ is the Poincaré dual to the cohomology class of the flux $c = \psi(\mu)$ (for a unique invariant measure μ) and the 1-dimensional subspace L_1 is spanned by λ. For any $j = 1, \ldots, g-1$ one has

$$\limsup_{n\to\infty} \frac{\log d(\gamma_n, L_j)}{\log n} = \nu_{j+1}$$

and

$$\frac{d(\gamma_n, L_g)}{\log n} = O(1),$$

where the implied constant depends only on X and on the choice of the Euclidean structure inducing the distance d in the homology space. The numbers $2, 1 + \nu_2$, $\ldots, 1 + \nu_g$ are the top g Lyapunov exponents *of the Teichmüller geodesic flow on $\mathcal{H}$ (with respect to $\mu_\mathcal{H}$)*.

Zorich's conjectures. We note that the inequality $1 > \nu_2$ follows from the work of Veech [**163**, Theorem 0.2]. Before we discuss general notions of Lyapunov exponents (as introduced by Furstenberg-Kesten [**76**] and Oseledets [**138**]) in further detail in §5.4.4 below, we note that the numerical experiments suggested in fact that inequalities among the ν_i should be strict (and for many applications, including to speed of convergence in ergodicity of linear flows, and for analyzing properties of the Teichmüller flow, this strictness is crucial). Indeed, Zorich conjectured [**176**, Conjecture 1] that $\nu_g > 0$ and [**176**, Conjecture 2] that

$$1 > \nu_2 > \cdots > \nu_{g-1} > \nu_g > 0.$$

An important consequence of $\nu_g > 0$ would be that L_g is Lagrangian, and simplicity would imply that the flag is Lagrangian.

5.4.2. Forni's theorems. The first of these conjectures was resolved in a pioneering work of Forni [**69**], who, building on connections to Hodge theory suggested in Kontsevich-Zorich [**108**, **109**] and developing a functional analytic theory of *invariant distributions* for linear flows on translation surfaces, proved (among several other important results) the first conjecture.

Theorem 5.4.2 ([**69**, Theorem 0.2]). *With notation as above,*

$$1 > \nu_2 \geq \nu_3 \geq \cdots \geq \nu_g > 0.$$

Ergodic integrals. Forni's techniques in fact proved a more general conjecture of Kontsevich-Zorich on ergodic integrals of smooth functions, which are crucial in applications.

Hyperbolicity. Theorem 5.4.2 (in particular that $\nu_2 < 1$) implies the *non-uniform hyperbolicity* of the Teichmüller geodesic flow on all strata (which, as mentioned above, had been shown by Veech [**163**] for all strata of quadratic differentials).

Other measures. In fact, Forni [**69**, Theorem 0.2′] extended the nonuniform hyperbolicity of the Teichmüller flow to *all* geodesic flow ergodic invariant probability measures, an important strengthening for further applications.

Zero exponents. Forni [**69**, §4] identified the *determinant locus*, which provides a mechanism for producing 0 Lyapunov exponents, which are important (for example) in producing examples of linear flows on translation surfaces with controlled deviation behavior. This and subsequent developments (see, for example, work of Aulicino [**14**–**16**] and Möller [**130**]) led to a conjecture of Forni-Matheus-Zorich [**72**] classifying mechanisms for 0 exponents, which was resolved (in a slightly modified form) by Filip [**67**].

5.4.3. Simplicity. Zorich's simplicity conjecture (that is, that all the inequalities between exponents are strict) was resolved by Avila-Viana [**23**], who used powerful general methods from dynamical systems (in particular, developing a notion of *pinching and twisting* cocycles) to prove simplicity of exponents for cocycles in some generality.

5.4.4. Lyapunov exponents. As mentioned in the introduction to this section, the proofs of the theorems we state here are well beyond the scope of this book. However, we would like to give the interested reader a (very) brief description of Lyapunov exponents in some generality, as well as a discussion of the Kontsevich-Zorich cocycle and its connection to Teichmüller flow. Broadly speaking, Lyapunov exponents are numerical invariants associated to ergodic transformations, flows, and actions which represent the asymptotic exponential growth rates of the eigenvalues of the derivative cocycle (see, for example, the supplement by Katok-Mendoza in the excellent reference volume of Katok-Hasselblatt [**98**] for a precise definition). For the initial part of our discussion, we follow Furman's exposition [**74**] and suggest the reader consult Filip's notes [**68**].

Random matrix products. As with many important concepts in dynamical systems, the idea of Lyapunov exponents emerged from probability theory. Just as the Birkhoff Ergodic Theorem 4.2.4 can be viewed as a generalization of the *strong law of large numbers* for sums of (finite-mean) independent random variables, the general theory of Lyapunov exponents starts with *products* of independently chosen random matrices, attempting to understand their *exponential growth rate*. Since products of randomly chosen numbers are well understood, we start with matrices with determinant 1. Fix μ to be a probability measure on $G = SL(d, \mathbb{R})$, and assume it satisfies a *finite first moment* condition

$$(5.4.1) \qquad \int_G \log \|g\| d\mu(g) < \infty,$$

where $\|\cdot\|$ is any norm on G (for concreteness, we can consider the operator norm with respect to the L^2-norm on $\mathbb{R}^d$). Let $\mathbf{P} = \mathbf{P}_\mu$ denote the product measure $\mu^\mathbb{N}$ on the space $\Omega = G^\mathbb{N}$. That is, a choice of a sequence $\{X_n \in G : n \in \mathbb{N}\}$ from $\mathbf{P}$ is a way of selecting independent, identically distributed (IID) matrices X_n from G using μ. Then we have the following classical theorem of Furstenberg-Kesten [76] (our version of the statement is from Furman [74, Proposition 1.1]):

Theorem 5.4.3. *Let $\{X_n \in G : n \in \mathbb{N}\}$ be a sequence of IID μ-matrices. Then with $\mathbf{P}$ probability 1, the limit*

$$(5.4.2) \qquad \lim_{N \to \infty} \frac{1}{N} \log \|X_N X_{N-1} \cdots X_1\|$$

exists, and its value $\lambda_1(\mu)$ is $\mathbf{P}$-almost everywhere constant and can be expressed as

$$\lambda_1(\mu) = \lim_{N \to \infty} \frac{1}{N} \int_G \log \|g\| d\mu^n(g) = \inf_{N > 0} \frac{1}{N} \int_G \log \|g\| d\mu^n(g),$$

where $\mu^n(g)$ is the n-fold convolution power of μ (that is, the push forward of the product measure on G^n under the map $(g_n, \ldots, g_1) \mapsto g_n g_{n-1} \cdots g_1$).

Kingman's subadditive ergodic theorem. The now standard proof of Theorem 5.4.3 uses *Kingman's subadditive ergodic theorem* [106, Theorem 1]. We state the theorem following Lalley [110] (and we refer the interested reader to the same reference for an elegant proof).

Theorem 5.4.4 ([110, Theorem 1]). *Let $(\Omega, \mathbf{P})$ be a probability measure space, and let T be a measure-preserving transformation. Suppose $\{g_n\}_{n \geq 1} \subset L^1(\Omega, \mathbf{P})$ satisfies the subadditivity relation*

$$g_{m+n}(\omega) \leq g_n(\omega) + g_m(T^n(\omega)).$$

Then for $\mathbf{P}$-almost every $\omega \in \Omega$,

$$\lim_{N \to \infty} \frac{g_N(\omega)}{N} = \gamma := \inf_{k \geq 1} \frac{1}{k} \int_\Omega g_k d\mathbf{P}.$$

Exercise 5.3. *Prove Theorem 5.4.3 using Theorem 5.4.4 applied to the shift map on the space $\Omega = G^\mathbb{N}$. See Furman [74, Proof of Proposition 1.1] for details.*

Subadditive sequences. In fact, many of the key ideas in the proof of Theorem 5.4.4 are contained in the following:

Exercise 5.4. *Let $\{a_n\}_{n \in \mathbb{N}}$ be a sequence of nonnegative real numbers satisfying the subadditivity condition*

$$a_{n+m} \leq a_n + a_m.$$

Show that

$$\lim_{n \to \infty} \frac{a_n}{n}$$

5.4. Quantitative renormalization and ergodicity

exists and is equal to

$$\gamma = \inf_{k \geq 1} \frac{a_k}{k}.$$

Exterior powers. Theorem 5.4.3 gives the existence of *one* exponent, the top rate of growth. For further exponents, one considers the action on *exterior powers* and defines the exponents inductively. Following Furman [**74**, page 937], we define, for $p = 1, \ldots, d$,

$$\bigwedge^p S_N = \bigwedge^p X_N \ldots \bigwedge^p X_1$$

(so $S_N = X_N \ldots X_1$ is the product) and we set $\lambda_1(\mu) + \cdots + \lambda_p(\mu)$ to be the **P**-almost sure limit of

$$\lim_{N \to \infty} \frac{1}{N} \log \left\| \bigwedge^p S_N \right\|.$$

Note that for any $A \in SL(d, \mathbb{R})$ we have $\bigwedge^d A = \det A = 1$, so the exponents $\lambda_i = \lambda_i(\mu)$ must satisfy

$$\lambda_1 + \cdots + \lambda_d = 0.$$

More concretely, if we write the product $S_N = X_N \ldots X_1$ in polar form (see [**74**, equation (1.3)])

$$S_N = V_N \operatorname{diag}(e^{a_1(N)}, \ldots, e^{a_d(N)}) U_N,$$

with $U_N, V_N \in SO(d)$ and $a_1 \geq a_2 \geq \cdots \geq a_k$, we have

$$\lim_{N \to \infty} a_p(N)/N = \lambda_p$$

for $p = 1, \ldots, d$.

Oseledets's theorem. Next, we state an important generalization by Oseledets [**138**] of the Furstenberg-Kesten result Theorem 5.4.3 to the setting of cocycles over ergodic transformations. A rough analogy is as follows: the Birkhoff theorem is to the strong law of large numbers as the Oseledets's theorem is to the Furstenberg-Kesten result. We follow the exposition of Filip [**68**], referring the interested reader there for more details. Given a probability space $(\Omega, \mathbf{P})$ and an ergodic measure-preserving transformation $F : \Omega \to \Omega$, a (measurable, continuous, smooth) *cocycle* over F is a measurable vector bundle $V \to \Omega$ equipped with (invertible) linear maps between fibers

$$F_\omega : V_\omega \to V_{F\omega}$$

and these maps vary (measurably, continuously, smoothly) with $\omega \in \Omega$. We require $F_\omega^2 : V_\omega \to V_{F^2\omega}$ to satisfy

(5.4.3) $$F_\omega^2 = F_{F\omega} \circ F_\omega.$$

The classical example of a cocycle is the *derivative cocycle* over a (volume-preserving) ergodic diffeomorphism F of a smooth manifold M, where the vector bundle is the tangent bundle TM and the associated maps are the derivative maps

$$DF_\omega : T_\omega M \to T_{F\omega}M.$$

The above equation (5.4.3) in this case is the chain rule. The random matrix products discussed above can be put into a cocycle framework by considering the cocycle over the shift map F on $\Omega = G^\mathbb{N}$, where the fiber over each point is $\mathbb{R}^d$. The theorem of Oseledets extends the Furstenberg-Kesten result to the growth of general cocycles. Let V be a (measurable) cocycle over an ergodic measure-preserving map $F : (\Omega, \mathbf{P}) \to (\Omega, \mathbf{P})$. Assume further that V is equipped on each fiber V_ω with a metric $\|\cdot\|_\omega$ such that

$$\int_\Omega \log^+ \|F_\omega\|_{op} d\mathbf{P}(\omega) < \infty,$$

where $\log^+(t) = \max(0, \log t)$ and $\|\cdot\|_{op}$ is the operator norm of the map F_ω as a map between the normed vector spaces V_ω and $V_{F\omega}$. Then we have the Oseledets *mulitplicative ergodic theorem*:

Theorem 5.4.5 ([**68**, Theorem 1.4]). *There exist real numbers $\lambda_1 > \lambda_2 > \cdots > \lambda_k$ and F-invariant subbundles $V^{\leq \lambda_i}$, $i = 1, \ldots, k$, of V defined for $\mathbf{P}$-almost every $\omega \in \Omega$ with*

$$0 \subsetneq V^{\leq \lambda_k} \subsetneq \cdots \subsetneq V^{\leq \lambda_1} = V$$

such that for vectors $v \in V_\omega^{\leq \lambda_i} \setminus V_\omega^{\lambda_{i+1}}$ we have

$$\lim_{N \to \infty} \frac{1}{N} \log \|F_\omega^N v\|_{F^N \omega} = \lambda_i.$$

Multiplicities. The multiplicity of an exponent λ_i is defined to be $(\dim V^{\leq \lambda_i} - \dim V^{\leq \lambda_{i+1}})$. We will in general write exponents as repeated with their appropriate multiplicity.

The Kontsevich-Zorich cocycle. How then does this relate to our situation? While we stated the Oseledets theorem for transformations, it is clear how to generalize it to flows (and possible to generalize to cocycles over general group actions). As such, we can consider the *derivative* cocycle for the Teichmüller flow g_t acting on a (connected component) of a stratum $\mathcal{H}$. Since we have local coordinates on strata which identify neighborhoods of $\mathcal{H}$ with the relative cohomology $H^1(X, \Sigma, \mathbb{R})$, which is a linear space, the *tangent space* to $\mathcal{H}$ at a point $\omega \in \mathcal{H}$ can also be identified with $H^1(X, \Sigma, \mathbb{R})$. To define the closely related *Kontsevich-Zorich cocycle*, we follow the original paper of Kontsevich-Zorich [**108**], [**109**, §5]: define the vector bundle V over $\mathcal{H}$ so that the fiber at a point $\omega \in \mathcal{H}$ is the *absolute* cohomology group $H^1(X, \mathbb{R})$. To lift the action of g_t to this bundle, we are technically using the natural flat connection, known as the Gauss-Manin connection (see [**108**,**109**] or Forni [**69**, §1] and the references within for further details).

Symplectic cocycles. Applying the Multiplicative Ergodic Theorem 5.4.5 to the action of g_t on this bundle gives a *symmetric* collection of Lyapunov exponents

$$1 = \lambda_1 \geq \lambda_2 \geq \cdots \geq \lambda_g \geq \lambda_{g+1} = -\lambda_g \geq \cdots \geq \lambda_{2g} = -\lambda_1.$$

The symmetry is due to the fact that this cocycle is *symplectic*. Morally speaking, what the cocycle is doing is keeping track of a geodesic trajectory in the universal cover over of the stratum $\mathcal{H}$. When a g_t trajectory leaves a fundamental domain for $\mathcal{H}$, it is brought back by an element of the mapping class group, and we are recording the action of this element on cohomology, which preserves the symplectic form given by intersection number. For a lucid and more pictorial introduction to this cocycle and its connection to interval exchange transformations, we refer the reader again to Zorich's survey [**177**]. We note that the exponents of the (derivative cocycle) for the Teichmüller flow are given by adding ±1 to the exponents λ_i for the Kontsevich-Zorich cocycle.

Continued fractions. In the genus 1 setting, there is a connection between this cocycle over the geodesic flow on the moduli space $SL(2, \mathbb{R})/SL(2, \mathbb{Z})$ of unit-area differentials on flat tori and continued fractions. See, for example, Zorich [**177**, §5.9] and Arnoux [**3**].

Connections to Hodge theory. There are many further connections between Lyapunov exponents and important features of moduli space, in particular relationships between the *sums* $\lambda_1 + \cdots + \lambda_g$ and *volumes* of strata $\mathcal{H}$. A crucial result connecting these is that of Eskin-Kontsevich-Zorich [**54**] (see also §6.7 below), which gives an explicit expression for this sum in terms of the orders of zeros of the stratum and a *Siegel-Veech* constant which governs the asymptotic growth of cylinders weighted by area. We refer the interested reader to, for example, Zorich [**177**, §5.8] and the references within.

5.4.5. Further applications. There are many further applications of the quantitative renormalization technology we have described above to the ergodic theory of billiards, linear flows, and interval exchange maps. We briefly highlight some examples of these results.

Weak mixing. As we discussed in §4.5.1, interval exchange transformations are *never* strong mixing, and rotations are not weak mixing. However, using quantitative renormalization ideas, Avila-Forni [**19**] showed that for IETs that are not rotations, they are almost surely weak mixing (and they showed similar results for flows). We note that unlike ergodicity, mixing and weak mixing of a flow and a first return map are not equivalent.

Theorem 5.4.6. *For any permutation π on d letters that is not a rotation almost every IET (with respect to Lebesgue measure on the simplex Δ) is weak mixing. For any component $\mathcal{H}$ of any stratum other than the strata of tori, $\mathcal{H}(\emptyset)$, for $\mu_{\mathcal{H}}$-almost every ω and Lebesgue almost-every θ, the flow ϕ_t^θ is weak mixing.*

Quantitative weak mixing. This result has been made quantitative by recent work of Avila-Forni-Safaee [20].

Billiards. Theorem 5.1.3 (and its quantitative strengthenings described above) holds for *almost every* translation surface $\omega \in \mathcal{H}$. However, these theorems do not say anything about an individual translation surface, and in particular, they do not apply to billiards since the set of surfaces which arise as unfoldings of billiards is a set of $\mu_{\mathcal{H}}$-measure 0. (This is true since the unfolding implies there are linear relationships between homologically independent saddle connections.) Theorem 5.1.3 was strengthened by Kerckhoff-Masur-Smillie [103], who, motivated by billiard flows and other low-dimensional dynamical systems, studied the set of nonuniquely ergodic directions on every ω. Given $\omega \in \mathcal{H}$, define

$$\text{NUE}(\omega) = \{\theta \in [0, 2\pi) : \phi_t^\theta \text{ is not uniquely ergodic}\}.$$

Theorem 5.4.7 ([103]). *For any $\omega \in \mathcal{H}$, NUE(ω) has Lebesgue measure 0. That is, for any translation surface ω and almost every direction θ, the linear flow ϕ_t^θ is uniquely ergodic, and in particular, every trajectory equidistributes with respect to Lebesgue measure on ω.*

More quantitative results. In particular, Theorem 5.4.7 implies that for any rational billiard table P and almost every direction θ, the billiard flow in direction θ equidistributes with respect to Lebesgue measure on P. Using quantitative renormalization ideas, Theorem 5.4.7 was strengthened quantitatively by Athreya-Forni [13], who showed that for any $\omega \in \mathcal{H}$ there is a constant $0 < \alpha = \alpha(\omega) < 1$, so that for any sufficiently regular mean-0 function f on X, for almost every θ, for all $x \in X$

$$\int_0^T f\left(\phi_t^\theta(x)\right) dt = O(T^\alpha),$$

where the implied constant depends on θ and f.

Weak mixing for billiards. Weak mixing in almost every direction is an *open* problem for general billiards in rational polygons. For *regular polygons* which do not tile the plane, this was shown by Avila-Delecroix [17, Theorem 1], who in fact showed it for all nonarithmetic lattice surfaces (see Chapter 7 for a precise definition) and even controlled the Hausdorff dimension of the set of nonweak mixing directions.

Wind-trees. A famous statistical physics model is the *Ehrenfest wind-tree* model, introduced by Paul and Tatiana Ehrenfest [50]. This consists of a $\mathbb{Z}^2$-array of periodic rectangular scatterers and a billiard dynamics within this array (that is a point mass moving at unit speed colliding elastically with the scatterers). Using the theory of Lyapunov exponents for the Teichmüller flow, Delecroix-Hubert-Lelièvre [45] showed that, *independent* of the size of the scatterers (as long as they do not fill up the plane), for almost every initial direction of travel, the *diffusion rate* is 2/3; that is, if $d(x, t, \theta)$ denotes the distance from the starting point x after time t for travel in direction θ, then for almost all θ,

$$\limsup_{t \to \infty} \frac{\log d(x, t, \theta)}{\log t} = \frac{2}{3}.$$

The number 2/3 is computed using the formula of Eskin-Kontsevich-Zorich [54] which we mentioned at the end of the previous subsection. For further details, and more recent developments see the paper of Delecroix-Zorich [46].

5.5. Nonunique ergodicity and Hausdorff dimension

We conclude this chapter with a discussion (without proofs) of some results on finer notions (beyond measure) of size of the set of nonuniquely ergodic directional flows.

5.5.1. Hausdorff dimension of nonergodic directions. Theorem 5.4.7 says that for any ω the set of nonuniquely ergodic directions has Lebesgue measure 0, which naturally leads to the issue of Hausdorff dimension of this set. Note that for the square torus, any nonuniquely ergodic direction is nonminimal and is thus rational, and the set of rational directions is countable. Any countable set has 0 Hausdorff dimension. We will see below in Chapter 7 that, like the torus, there are surfaces which continue to have only a countable collection of nonergodic directions, but for now we discuss the more complicated situation of general surfaces. Masur [119] proved a universal upper bound for the Hausdorff dimension of the set $\mathrm{NUE}(\omega)$.

Theorem 5.5.1 ([119]). *For any $\omega \in \mathcal{H}$*

$$\mathrm{HDim}(\mathrm{NUE}(\omega)) \leq \frac{1}{2}.$$

Almost sure Hausdorff dimension. Using the dynamics of the $SL(2, \mathbb{R})$-action, Masur-Smillie [121] showed that the function $\mathrm{HDim}(\mathrm{NUE}(\omega))$ is almost everywhere constant on $\mathcal{H}$.

Theorem 5.5.2 ([121]). *For any connected component of a stratum $\mathcal{H}$ other than $\mathcal{H}(\emptyset)$, there is a constant $\frac{1}{2} \geq d = d_{\mathcal{H}} > 0$ depending only the component such that for $\mu_{\mathcal{H}}$-almost all $\omega \in \mathcal{H}$*

$$\mathrm{HDim}(\mathrm{NUE}(\omega)) = d.$$

The value of $d_{\mathcal{H}}$. The question then arose as to the value of $d_{\mathcal{H}}$ for various $\mathcal{H}$. Athreya-Chaika [8] showed that:

Theorem 5.5.3. *The constant $d_{\mathcal{H}(2)}$ for the stratum $\mathcal{H}(2)$ of genus 2 surfaces with one double zero is 1/2.*

Higher genus. Theorem 5.5.3 was extended by Chaika-Masur [32] to hyperelliptic components in higher genus. Recall (§3.5.1) that the strata $\mathcal{H}(2g-2)$ and $\mathcal{H}(g-1, g-1)$ each have a *hyperelliptic component*, consisting of ω, so that the underlying Riemann surface X has an involution $\iota : X \to X$ (i.e., a biholomorphic map with $\iota^2 = \text{Id}$) such that

$$\iota_* \omega = -\omega.$$

Theorem 5.5.4. *Let $g \geq 2$. Let $\mathcal{H}$ be a hyperelliptic component of either $\mathcal{H}(2g-2)$ and $\mathcal{H}(g-1, g-1)$. Then $d_{\mathcal{H}} = \frac{1}{2}$.*

Connections to IETs. Both Theorem 5.5.3 and Theorem 5.5.4 are proved by an analysis of an appropriate related family of interval exchange maps and by exploiting a construction of Keane [101] of minimal, nonuniquely ergodic IETs.

The Veech example. As we discussed in §4.3.2, there are explicit results about Hausdorff dimension of the set of nonuniquely ergodic flows arising from the Veech slit torus example. Recall that these are the translation surfaces ω given in Figure 2.5, which are square tori glued along a vertical slit of length α.

Diophantine conditions. The length of the slit $\alpha \in \mathbb{R}$ is said to be *Diophantine* if there are constants $c, s > 0$ such that

$$\left| \alpha - \frac{p}{q} \right| \geq c q^{2+s}$$

for all $\frac{p}{q} \in \mathbb{Q}$. Cheung [36] showed that if the height of the slit is Diophantine, the Hausdorff dimension of the set of nonuniquely ergodic directions is as large as possible.

Theorem 5.5.5 ([36]). *If ω_α is the genus 2 translation surface constructed from square tori by gluing along a vertical slit of length α and α is Diophantine, then*

$$\text{HDim}(\text{NUE}(\omega)) = \frac{1}{2}.$$

Dichotomy for $\text{HDim}(\text{NUE}(\omega))$. Theorem 5.5.5 was extended by Cheung-Hubert-Masur [37] to a dichotomy for this family of surfaces. They showed that the only possibilities for $\text{HDim}(\text{NUE}(\omega_\alpha))$ are 1/2 or 0, and which case we fall into depends on the continued fraction expansion of α.

5.5. Nonunique ergodicity and Hausdorff dimension

Theorem 5.5.6 ([37]). *Let q_k be the sequence of denominators in the continued fraction expansion of α. Then $\mathrm{HDim}(\mathrm{NUE}(\omega)) = 0$ or $\frac{1}{2}$, with the latter case occurring if and only if α is irrational and*

$$\sum_{k=1}^{\infty} \frac{\log \log q_{k+1}}{q_k} < \infty.$$

Chapter 6

Counting and Equidistribution

In the previous chapter, we saw how the recurrence behavior of an orbit of a translation surface ω under the positive diagonal subgroup of $SL(2,\mathbb{R})$ gives information about the ergodic and equidistribution properties of the vertical flow $v_t \to \phi_t$. In this chapter, we will follow a strategy of Veech [**165**] and Eskin-Masur [**57**], inspired by earlier work of Eskin-Margulis-Mozes [**55**], to demonstrate how other parts of the $SL(2,\mathbb{R})$-action can be used to obtain precise asymptotic information about *counting problems* on translation surfaces. We first discuss classical lattice point counting in §6.1, then state our main results in §6.2. In §6.3, we introduce the notion of the *Siegel-Veech transform*, which collates information about saddle connections on translation surfaces by associating to each (bounded, compactly supported) function on $\mathbb{C}$ a function on the space of translation surfaces. We then show how to go from counting to dynamics in §6.4 and then prove our main dynamical result in §6.5. We discuss further counting results in §6.6 and then discuss how counting asymptotics are related to the volumes of moduli spaces in §6.7. Our exposition is strongly influenced by Eskin's survey [**53**].

6.1. Lattice point counting

We start by recalling our discussion from the introduction (specifically, §1.4) on the simplest translation surface, the square flat torus $(\mathbb{C}/\mathbb{Z}[i], dz)$. We showed there that counting primitive closed geodesics passing through 0 of length at most R was equivalent to understanding the primitive lattice point counting function

$$N_{\text{prim}}(R) = \#\{m + ni \in \mathbb{Z}[i] : \gcd(m,n) = 1, m^2 + n^2 \leq R^2\}$$

and that $N_{\text{prim}}(R)$ was asympotic to $\frac{1}{\zeta(2)}\pi R^2$.

6.1.1. The space of lattices. As we discussed in Chapter 1, the flat torus $\mathbb{C}/\mathbb{Z}[i]$ has multiple generalizations, and before turning to translation surfaces, we first discuss the space X_n of *unimodular lattices* in $\mathbb{R}^n$, for $n \geq 2$. Recall that a *unimodular lattice* Λ in $\mathbb{R}^n$ is a discrete covolume 1 additive subgroup of $\mathbb{R}^n$. The space X_n can be identified with the homogeneous space $SL(n,\mathbb{R})/SL(n,\mathbb{Z})$ via the map

$$gSL(n,\mathbb{Z}) \mapsto g\mathbb{Z}^n.$$

Haar measure. We have a natural measure μ_n on X_n inherited from Haar measure on $SL(n,\mathbb{R})$, which we define as follows: for a Borel subset $A \subset SL(n,\mathbb{R})$, we define the Haar measure μ_n of A to be the Lebesgue measure of the cone over A; that is,

$$\mu_n(A) = m(\{tg : g \in A, 0 \leq t \leq 1\}),$$

where m denotes Lebesgue measure on $M_n(\mathbb{R}) \cong \mathbb{R}^{n^2}$. Note the similarity to the definition of the MSV measure on strata $\mathcal{H}$ of unit-area translation surfaces, with area in place of determinant. In that construction, we used Lebesgue measure in local coordinates (determined by relative cohomology) on Ω and the analogous cone construction to define a measure on the area 1 hypersurface $\mathcal{H}$.

Primitivity. Given a lattice Λ, we say $v \in \Lambda$ is *primitive* if it is not an integer multiple of another vector $w \in \Lambda$ other than a multiple of ± 1. Equivalently, if $\Lambda = g\mathbb{Z}^n$, the set Λ_{prim} of primitive vectors is

$$\Lambda_{\text{prim}} = g\mathbb{Z}^n_{\text{prim}},$$

where

$$\mathbb{Z}^n_{\text{prim}} = \{(v_1,\ldots,v_n)^T \in \mathbb{Z}^n : \gcd(v_1,\ldots,v_n) = 1\} = SL(n,\mathbb{Z}) \cdot e_1$$

and e_1 is the first standard basis vector in $\mathbb{R}^n$. We have the following higher-dimensional generalization of Exercise 1.5.

Exercise 6.1. *Prove*

$$\lim_{R \to \infty} \frac{N_{\text{prim}}(R)}{\text{vol}(B(0,R))} = \frac{1}{\zeta(n)}.$$

6.1. Lattice point counting

Siegel's formula. Let $B_c(\mathbb{R}^n)$ be the set of bounded compactly supported functions on $\mathbb{R}^n$. For $f \in B_c(\mathbb{R}^n)$ and $\Lambda \in X_n$, define
$$\widehat{f}(\Lambda) = \sum_{v \in \Lambda_{\text{prim}}} f(v).$$

Let ρ_n be the scaling of μ_n so that $\rho_n(X_n) = 1$. We refer to ρ_n as the Haar probability measure on X_n. Then we have [152, page 341] *Siegel's integral formula:*

$$(6.1.1) \qquad \int_{X_n} \widehat{f}(\Lambda) d\rho_n(\Lambda) = \frac{1}{\zeta(n)} \int_{\mathbb{R}^n} f(x) dx.$$

Linear functionals. We indicate a proof of (6.1.1) which will contain the key idea for our generalization of this to the Siegel-Veech formula on the space of translation surfaces and which uses the Riesz representation theorem (see, for example, [143, §2]). We first need the following:

Lemma 6.1.1. *For $f \in B_c(\mathbb{R}^n)$, one has $\widehat{f} \in L^1(X_n, \rho_n)$. In fact, $\widehat{f} \in L^p(X_n, \rho_n)$ for any $p < n$.*

Proof. It suffices to show this for $f = \mathbf{1}_{B(0,1)}$. For T sufficiently large
$$\rho_n(\Lambda : \widehat{f}(\Lambda) > T) \leq c_n T^{-n},$$
where $c_n = \sum_{1 \leq k < n} \text{vol}(B(0,1))^k$. This inequality follows from the fact that if a lattice has more than T primitive vectors of length at most 1, it must have, for some $1 \leq k < n$, k basis vectors of length at most $1/T^{1/k}$, since primitive vectors are linear combinations (with coprime integer coefficients) of basis vectors. Thus we can write $\Lambda = g\mathbb{Z}^n$, where k of the columns of g have length at most $1/T^{1/k}$, and therefore, by the construction of the measure ρ_n on X_n,

$$\rho_n\left(\Lambda : \widehat{f}(\Lambda) > T\right) \leq \sum_{k=1}^{n-1} \text{vol}(B(0, T^{-1/k}))^k$$
$$\leq \sum_{k=1}^{n-1} \text{vol}(B(0,1))^k (T^{-n/k})^k$$
$$= c_n T^{-n}. \qquad \square$$

To prove (6.1.1) now note that the map $T : B_c(\mathbb{R}^n) \to \mathbb{R}^+$ given by
$$T(f) = \int_{X_n} \widehat{f} d\rho_n$$
is an $SL(n, \mathbb{R})$-invariant linear functional. That is, for $g \in SL(n, \mathbb{R})$,
$$T(f \circ g) = T(f),$$
and for $a, b \in \mathbb{R}$ and $f, h \in B_c(\mathbb{R}^n)$,
$$T(af + bh) = aT(f) + bT(h).$$

By the Riesz representation theorem (see, for example, [143, §2]), linear functionals on $C_c(\mathbb{R}^n) \subset B_c(\mathbb{R}^n)$ arise from integration against measures on $\mathbb{R}^n$, and by $SL(n, \mathbb{R})$-invariance of T, there exists an $SL(n, \mathbb{R})$-invariant measure m_T on $\mathbb{R}^n$ such that

$$T(f) = \int_{\mathbb{R}^n} f \, dm_T.$$

Exercise 6.2. *Assuming uniqueness (up to scaling) of Haar measure on $SL(n, \mathbb{R})$, show that any $SL(n, \mathbb{R})$-invariant Borel measure ν on $\mathbb{R}^n$ is of the form*

$$\nu = a\delta_0 + bm,$$

where m is Lebesgue measure on $\mathbb{R}^n$ and a, b are nonnegative constants. Hint: Use the fact that $SL(n, \mathbb{R})$ acts transitively on $\mathbb{R}^n \setminus \{0\}$.

By the exercise, we have that

$$m_T = a\delta_0 + bm.$$

To determine these constants, we follow an argument as outlined in [53]. First, by considering $f = \mathbf{1}_{B(0,R)}$ as $R \to 0$, we see that $a = 0$, since $T(f) \to 0$. Considering $R \to \infty$ and using Exercise 6.1, we see that $b = \frac{1}{\zeta(n)}$.

Counting, volumes, and induction. An application of Siegel's formula and these counting results, which serves as a model for similar types of computations in our setting of the moduli space of translation surfaces, is the computation of the *natural* volume of the spaces X_n, namely $\mu_n(X_n)$ where μ_n is defined in terms of the Lebesgue measure on $M_n(\mathbb{R})$ normalized so $M_n(\mathbb{Z})$ has unit covolume. By applying the integral formula (6.1.1) to the indicator function $\mathbf{1}_{B(0,\epsilon)}$ of the ball of radius ϵ centered at $0 \in \mathbb{R}^n$ and analyzing the resulting integrals as $\epsilon \to 0$, Siegel [153, Lecture XV] showed the following elegant induction:

(6.1.2) $$n\mu_n(X_n) = \zeta(n)(n-1)\mu_{n-1}(X_{n-1}).$$

Combining this with a direct computation from hyperbolic geometry showing that in the base case $n = 2$, $\mu_2(X_2) = \frac{1}{2}\zeta(2)$, Siegel concluded the following formula for the volume $\mu_n(X_n)$:

(6.1.3) $$\mu_n(X_n) = \frac{1}{n}\zeta(n)\zeta(n-1)\ldots\zeta(2).$$

Degenerations. The $\epsilon \to 0$ limit alluded to above involves understanding complements of compact sets in X_n. Mahler [114] showed a *compactness criterion* which showed that a subset $A \subset X_n$ is precompact if and only if there is a $\epsilon > 0$ so that for every lattice in A, the shortest nonzero vector has length at least ϵ. The idea behind the induction is that if we indeed do have a primitive lattice vector of length at most ϵ, typically there will be a complementary sublattice of dimension $n-1$ and covolume $\frac{1}{\epsilon}$; that is, the neighborhood of the boundary of X_n can be understood as a product of a Euclidean ball in $\mathbb{R}^n$ and the space X_{n-1}, with appropriate normalizations. We will discuss a version of this argument for strata in more detail in §6.7.

6.2. Saddle connections and holonomy vectors

We recall some notation from §2.4. Given a translation surface $\omega \in \mathcal{H}$, a *saddle connection* γ on ω is a geodesic in the flat metric determined by ω that connects two zeros of ω, with none in its interior. That is, $\gamma : [0, L] \to X$, $\gamma(0), \gamma(L)$ are zeros of ω, and $\gamma(t)$ is not a zero of ω for $t \in (0, L)$. Here

$$L = \ell(\gamma) = \left|\int_\gamma \omega\right| = |z_\gamma|$$

is the length of γ in the flat metric determined by ω, and

$$z_\gamma = \int_\gamma \omega$$

is the *holonomy vector* of ω. Recall that $\mathrm{SC}(\omega)$ is the set of all saddle connections on ω and

$$\Lambda_\omega = \{z_\gamma : \gamma \in \mathrm{SC}(\omega)\} \subset \mathbb{C}.$$

We showed, in Lemma 2.4.7, that $\mathrm{SC}(\omega)$ is a countable set and Λ_ω is a countable discrete subset of $\mathbb{C}$. The discreteness of Λ_ω implies that the set

$$\Lambda_\omega(R) = \Lambda_\omega \cap B(0, R)$$

is finite for any $R > 0$.

Counting. We can then define the counting function

$$N(\omega, R) = \#\Lambda_\omega(R).$$

Motivated by the study of special trajectories for rational polygonal billiards, Masur [**117**] showed that for *any* translation surface ω, this function has quadratic growth: there are constants $0 < c_1 \leq c_2 < \infty$ such that

(6.2.1) $$c_1 R^2 \leq N(\omega, R) \leq c_2 R^2.$$

Note that if we have a rational polygonal billiard table P, billiard trajectories connecting two corners of P (known as *generalized diagonals*) lead to saddle connections on the unfolded surface ω_P.

6.2.1. Almost sure asymptotics.
The main theorem we will prove in this chapter is a result of Eskin-Masur [57], an almost sure asymptotic formula for $N(\omega, R)$ (which, it should be noted, does not immediately imply any such result for billiards, since the set of surfaces arising from billiards has measure 0):

Theorem 6.2.1. *Let $\mathcal{H}$ be a connected component of a stratum of translation surfaces. There is a constant $0 < c_{\mathcal{H}} < \infty$ such that for $\mu_{\mathcal{H}}$-almost every $\omega \in \mathcal{H}$,*

$$(6.2.2) \qquad \lim_{R \to \infty} \frac{N(\omega, R)}{\pi R^2} = c_{\mathcal{H}}.$$

Siegel-Veech constants. The constants $c_{\mathcal{H}}$, known as *Siegel-Veech constants*, can be computed for many strata; see, for example, [58]. This result is a key part of a continuing story of counting results, starting with the aforementioned bounds (6.2.1) of Masur. Veech [165] introduced the idea of the *Siegel-Veech* transform and showed an L^1-counting result, namely, that for $c_{\mathcal{H}}$ as above,

$$(6.2.3) \qquad \lim_{R \to \infty} \int_{\mathcal{H}} \left| \frac{N(\omega, R)}{\pi R^2} - c_{\mathcal{H}} \right| d\mu(\omega) = 0.$$

He proved this result via a strategy, inspired by work of Eskin-Margulis-Mozes [55] on the quantitative Oppenheim conjecture in number theory. We will follow this strategy, as developed further by Eskin-Masur [57], to prove Theorem 6.2.1.

6.3. Siegel-Veech transforms

We now define the *Siegel-Veech* transform. Let $B_c(\mathbb{C})$ denote the set of bounded, compactly supported functions on $\mathbb{C}$. For $f \in B_c(\mathbb{C})$ we define its *Siegel-Veech transform*, denoted by $\widehat{f}$ or f^{SV}, a function on a stratum $\mathcal{H}$, by

$$\widehat{f}(\omega) = f^{SV}(\omega) = \sum_{z \in \Lambda_\omega} f(z).$$

Theorem 6.3.1 ([165]). *Let $\rho_{\mathcal{H}}$ denote the Masur-Smillie-Veech measure normalized to be a probability measure; that is, $d\rho_{\mathcal{H}} = \frac{1}{\mu_{\mathcal{H}}(\mathcal{H})} d\mu_{\mathcal{H}}$. Then for $f \in B_c(\mathbb{C})$,*

$$(6.3.1) \qquad \int_{\mathcal{H}} \widehat{f} d\rho_{\mathcal{H}} = \frac{1}{\mu_{\mathcal{H}}(\mathcal{H})} \int_{\mathcal{H}} \widehat{f} d\mu_{\mathcal{H}} = c_{\mathcal{H}} \int_{\mathbb{C}^*} f(z) dm(z).$$

The constant $c_{\mathcal{H}}$ is the same constant which appears in (6.2.2), and $dm(z)$ is Lebesgure measure on $\mathbb{C}^*$.

Integrability. To prove this theorem, we will use the fundamental result of the integrability of $\widehat{f}$ (proved by Veech [165, Theorem 0.5]):

Black Box 6.3.2. *For each $f \in B_c(\mathbb{C})$, the Siegel-Veech transform $\widehat{f}$ is integrable; that is, $\widehat{f} \in L^1(\mathcal{H}, \mu)$.*

Cusp neighborhoods. The proof of Black Box 6.3.2 is via estimating the measure of the set of surfaces which have many short saddle connections, a condition that describes a neighborhood of the boundary in $\mathcal{H}$.

Linear functionals. We now show how to obtain Theorem 6.3.1 using Black Box 6.3.2, following the same strategy we followed to prove Siegel's formula (6.1.1) in the space of lattices. Since for $f \in B_c(\mathbb{C})$, $\widehat{f} \in L^1$

$$f \longmapsto \int_{\mathcal{H}} \widehat{f} d\mu$$

is an $SL(2, \mathbb{R})$-invariant linear functional on $B_c(\mathbb{C}^*)$ (since μ is $SL(2, \mathbb{R})$-invariant). By the Riesz representation theorem (see, for example, [143, §2]), we have that there is an $SL(2, \mathbb{R})$-invariant measure η on $\mathbb{C}^*$ so that

$$\int_{\mathcal{H}} \widehat{f} d\mu = \int_{\mathbb{C}} f(z) d\eta(z).$$

Up to scaling, the only $SL(2, \mathbb{R})$-invariant measures on $\mathbb{C}$ are δ-measure at the origin and Lebesgue measure on $\mathbb{C}^*$. Since for every translation surface, there is a lower bound on the shortest saddle connection and there are nonzero saddle connections for a positive measure set of translation surfaces, η is not the δ-measure at 0, and therefore, there is a $c = c_{\mathcal{H}}$ such that $d\eta(z) = cdm(z)$, yielding (6.3.1).

Further integrability properties. The proof above can be modified to show a version of Theorem 6.3.1 where μ is replaced by any $SL(2, \mathbb{R})$-invariant measure ν where $\widehat{f} \in L^1(\nu)$ for $f \in B_c(\mathbb{C})$. Using work of Avila-Eskin-Möller [18], this condition is satisfied by every ergodic $SL(2, \mathbb{R})$-invariant measure ν. Moreover, Athreya-Cheung-Masur [10] show that $\widehat{f} \in L^{2+\delta}(\mu)$ for an explicit $\delta > 0$ depending on the combinatorics of the stratum $\mathcal{H}$, and Athreya-Chaika [7] show that $\widehat{f} \notin L^3(\mu)$. Using work of Dozier [49], Bonnafoux [27, Theorem 1.6] has shown that for every $SL(2, \mathbb{R})$-invariant measure ν, there is a $\kappa > 0$ so that $\widehat{f} \in L^{2+\kappa}(\nu)$.

6.4. From counting to dynamics

In this section, we show how to transform the problem of counting saddle connections (or more precisely their holonomy vectors) to a problem on the dynamics of the $SL(2, \mathbb{R})$-action on strata. We first define a family of averaging operators which will be crucial in our construction.

6.4.1. Averaging operators.
Let X be an $SL(2, \mathbb{R})$ space (that is, a space equipped with an $SL(2, \mathbb{R})$-action). For our purposes, X will be either $\mathbb{C}$ with the $\mathbb{R}$-linear action or the stratum $\mathcal{H}$. We define the averaging operators A_t as

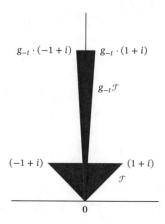

Figure 6.1. The triangle trick.

follows: given a function $h : X \to \mathbb{C}$, define

$$(6.4.1) \qquad (A_t h)(x) := \frac{1}{2\pi} \int_0^{2\pi} h(g_t r_\theta x) d\theta.$$

6.4.2. The triangle trick. We will show how to turn the problem of counting holonomy vectors of saddle connections on a *fixed* surface ω in varying regions $B(0, R)$ to a problem (an average) of counting holonomy vectors for *varying* surfaces in a fixed region, which will allow us to apply the ergodic theory of $SL(2, \mathbb{R})$ acting on $\mathcal{H}$ to our counting problem. We will do this by building a fixed set which captures the contributions of vectors of a bounded length on a fixed surface, when appropriately integrated over a family of surfaces. Let $\mathcal{T}$ be the triangle with vertices at $0, 1+i, -1+i$. Applying g_{-t} to $\mathcal{T}$ stretches the vertical and contracts the horizontal to yield a tall, thin triangle; see Figure 6.1. This triangle $g_{-t}\mathcal{T}$ has vertices at

$$g_{-t}.0 = 0, \quad g_{-t}.(1+i) = e^{-t/2} + ie^{t/2}, \quad g_{-t}.(-1+i) = -e^{-t/2} + ie^{t/2}.$$

For any $w \in g_{-t}\mathcal{T}$,

$$(6.4.2) \qquad |w| \leq \sqrt{e^t + e^{-t}}.$$

For $z \in \mathbb{C}$, define the set

$$\Theta_t(z) = \{\theta : g_t r_\theta z \in \mathcal{T}\}.$$

This is the set of angles that are needed to rotate the vector z so that it lies in the triangle $g_{-t}\mathcal{T}$. By (6.4.2), if $\Theta_t(z)$ is nonempty, $|z| < \sqrt{e^t + e^{-t}}$. Further, for $|z| \leq e^{t/2}$, if $z = |z|e^{i\theta_z}$, then setting

$$\psi_t = \arctan(e^{-t}),$$

6.4. From counting to dynamics

we claim
$$\Theta_t(z) = (\pi/2 - \theta_z - \psi_t, \pi/2 - \theta_z + \psi_t).$$
To see the claim note that rotating z by $\pi/2 - \theta_z$ puts it on the y-axis, and furthermore $\pm\psi_t$ are the angles that the sides of the triangle make with the y-axis, proving the claim. In particular, for any z such that $\Theta_t(z)$ is nonempty,
$$|\Theta_t(z)| \leq 2\psi_t.$$
Thus for any $z \in \mathbb{C}$,

(6.4.3) $$\frac{\psi_t}{\pi} \mathbf{1}_{B(0,e^{t/2})}(z) \leq (A_t \mathbf{1}_{\mathcal{T}})(z) \leq \frac{\psi_t}{\pi} \mathbf{1}_{B(0,\sqrt{e^t+e^{-t}})}(z).$$

Summing (6.4.3) over the set of holonomy vectors $z \in \Lambda_\omega$, we obtain, for any $t > 0$,

(6.4.4) $$\frac{\psi_t}{\pi} N(\omega, e^{t/2}) \leq A_t(\mathbf{1}_{\mathcal{T}})^{SV}(\omega) \leq \frac{\psi_t}{\pi} N(\omega, \sqrt{e^t + e^{-t}}).$$

We write
$$s(t) = t + \log(1 + e^{-2t}),$$
so
$$\sqrt{e^t + e^{-t}} = e^{t/2}\sqrt{1 + e^{-2t}} = e^{s(t)/2}.$$
We then have
$$\frac{1}{\psi_t} A_t(\mathbf{1}_{\mathcal{T}})^{SV}(\omega) \leq \frac{1}{\pi} N(\omega, e^{s(t)/2}).$$
Multiplying both sides by $\psi_{s(t)}$ and applying (6.4.4) to $s(t)$, we have

(6.4.5) $$\frac{\psi_{s(t)}}{\psi_t} A_t(\mathbf{1}_{\mathcal{T}})^{SV}(\omega) \leq \frac{\psi_{s(t)}}{\pi} N(\omega, e^{s(t)/2}) \leq A_{s(t)}(\mathbf{1}_{\mathcal{T}})^{SV}(\omega).$$

We note that
$$\lim_{t \to \infty} \frac{\psi_{s(t)}}{\psi_t} = \lim_{t \to \infty} \frac{\psi_t}{e^{-t}} = 1,$$
so putting $R = e^{s(t)/2}$, we have that

(6.4.6) $$\lim_{R \to \infty} \frac{N(\omega, R)}{\pi R^2} = \lim_{t \to \infty} A_{s(t)}(\mathbf{1}_{\mathcal{T}})^{SV}(\omega) = \lim_{t \to \infty} A_t(\mathbf{1}_{\mathcal{T}})^{SV}(\omega),$$

where this equality should be interpreted as saying that
$$\lim_{R \to \infty} \frac{N(\omega, R)}{\pi R^2}$$
exists if
$$\lim_{t \to \infty} A_t(\mathbf{1}_{\mathcal{T}})^{SV}(\omega)$$
does, in which case the two limits are equal.

Conclusion. Thus, we have reduced our problem of understanding the quadratic asymptotics of $N(\omega, R)$ to the limiting behavior of the integrals $A_t(\mathbf{1}_{\mathcal{T}})^{SV}(\omega)$.

6.5. Equidistribution of circles

The next step in proving Theorem 6.2.1 is now to understand the limits of the integrals $A_t(\mathbf{1}_{\mathcal{T}})^{SV}(\omega)$. The main result of this section is a corollary of a general ergodic theorem of Nevo [135] which tells you how to compute these limits. Our proof is a simplification of an argument made in Athreya-Fairchild-Masur [12].

Theorem 6.5.1. *Let $\varphi \in C_c(\mathbb{C})$. Then for $\rho_{\mathcal{H}}$-almost every $\omega \in \mathcal{H}$,*

$$(6.5.1) \qquad \lim_{t\to\infty} A_t(\widehat{\varphi})(\omega) = \int_{\mathcal{H}} \widehat{\varphi} \, d\rho_{\mathcal{H}} = c_{\mathcal{H}} \int_{\mathbb{C}} \varphi(z) dm.$$

Nevo's theorem. We will prove Theorem 6.5.1 modulo Nevo's theorem, which we state as a Black Box. We first make a definition.

K-finiteness. We say a function $f : \mathcal{H} \to \mathbb{C}$ is *K-finite* if after defining $f_\theta(\omega) := f(r_\theta \omega)$, the span of the functions $\{f_\theta(\omega) : \theta \in [0, 2\pi)\}$ is finite-dimensional. Equivalently, the function $p(\theta) = f_\theta(\omega)$ is a trigonometric polynomial of finite degree.

Black Box 6.5.2 ([135, Theorem 1.1]). *Suppose μ is an ergodic $SL(2,\mathbb{R})$-invariant probability measure on $\mathcal{H}$. Assume $f \in L^{1+\kappa}(\mathcal{H}, \mu)$ for some $\kappa > 0$ and that it is K-finite. Let $\eta \in C_c(\mathbb{R})$ be a continuous nonnegative bump function with compact support and of unit integral, which we refer to as a* mollifier. *Then for $\rho_{\mathcal{H}}$-almost every $\omega \in \mathcal{H}$,*

$$\lim_{t\to\infty} \int_{-\infty}^{\infty} \eta(t-s)(A_s f)(\omega) ds = \int_{\mathcal{H}} f \, d\rho_{\mathcal{H}}.$$

Approximation by K-finite functions. Our first step in proving Theorem 6.2.1 using Black Box 6.5.2 is to construct K-finite functions which sufficiently approximate $\widehat{\varphi}$, which we do by constructing a family of K-finite functions which are dense in the continuous functions.

The support. Given φ with compact support, define the rotation-invariant set $H = \overline{B(0, \ell_\varphi)}$, a closed ball with radius ℓ_φ chosen so that $\varphi \leq \|\varphi\|_\infty \mathbf{1}_H$. Define the slightly larger rotation-invariant set $H_1 = \overline{B(0, l_\varphi + 1)}$. The reason we introduce the set H_1 will be given in (6.5.2) and has to do with convolutions.

An algebra of functions. Consider the set $\mathcal{F} \subset C(H)$ given by

$$\mathcal{F} = \{f_{m,n} : m, n \in \mathbb{Z}\}$$

where for $z = re^{i\theta}$,

$$f_{m,n}(z) = r^m e^{in\theta}.$$

Define the subalgebra $\mathcal{A} \subset C(H)$ to be the $\mathbb{C}$-linear span of $\mathcal{F} \cup \{1\}$, the function that is identically 1 on H.

6.5. Equidistribution of circles

Exercise 6.3. *Prove that $\mathcal{F}$ is closed under multiplication and complex conjugation, and conclude that $\mathcal{A}$ is an algebra. Show further that the elements of $\mathcal{F}$ separate points; that is, for any pair of distinct points $z, w \in \mathbb{C}$, there is an $f \in \mathcal{F}$ so that*
$$f(z) \neq f(w).$$

Siegel-Veech transforms. We claim that Siegel-Veech transforms of functions in the algebra $\mathcal{A}$ are K-finite. This is a consequence of the following:

Exercise 6.4. *Prove that $f_{m,n}$ is K-finite for $f_{m,n} \in \mathcal{F}$ by showing*
$$(f_{m,n} \circ r_\theta)(z) = e^{in\theta} f_{m,n}(z).$$

The Stone-Weierstrass theorem. Exercise 6.4 implies that the $\mathbb{C}$-linear span of
$$\left\{ \widehat{f_\theta} : \theta \in [0, 2\pi) \right\}$$
is exactly the $\mathbb{C}$-linear span of $\widehat{f}$. By linearity, if $f \in \mathcal{A}$, both f and $\widehat{f}$ are K-finite. We recall the Stone-Weierstrass theorem (see, for example, [143, Theorem 7.26]):

Black Box 6.5.3 (Stone-Weierstrass). *Suppose X is a locally compact Hausdorff space and $\mathcal{A}$ is a subalgebra of $C_c(X)$. Then A is dense in $C_c(X)$ (given the topology of uniform convergence) if and only if it separates points and vanishes nowhere.*

Uniform convergence. We can conclude, using Black Box 6.5.3, that K-finite functions are dense in the uniform topology in $C(H)$. Now let $\{f_n\}_{n \in \mathbb{N}}$ be a sequence of K-finite functions which converge uniformly to φ, so $\text{Im}(f_n)$ converges uniformly to zero and $\text{Re}(f_n)$ converges uniformly to φ. By replacing f_n with $\text{Re}(f_n)$, we can assume that each f_n is real valued.

Mollifiers and scaling. We now fix our mollifier $\eta \in C_c^\infty(\mathbb{R})$, with $\eta(t) \geq 0$,
$$\int_{-\infty}^{\infty} \eta(t)\, dt = 1,$$
and support of η in $[-1, 1]$. For $f \in C_0^\infty(\mathbb{C})$ and $z \in \mathbb{C}$, denote the convolution by
$$(\eta * f)(z) := \int_{-\infty}^{\infty} \eta(t) f(g_{-t} z)\, dt.$$

For $\kappa > 0$, define the rescaling
$$\eta_\kappa(t) = \kappa^{-1} \eta(t/\kappa)$$
whose support is in $\kappa \cdot [-1, 1] = [-\kappa, \kappa]$. Note that
$$\lim_{\kappa \to 0} \eta_\kappa(t) = \delta_0(t)$$
where δ_0 is the Dirac-δ distribution at 0.

Convergence of convolutions. Note that as $\kappa \to 0$, the convolutions $\eta_\kappa * \varphi$ converge uniformly to φ. Indeed, for $\epsilon > 0$ choose κ_0 so that for $|t| < \kappa_0$,

$$|\varphi(g_{-t}z) - \varphi(z)| < \epsilon$$

and so that for any $\kappa < \kappa_0$ and $z \in H$, $g_{-\kappa}z \in H_1$, so

(6.5.2) $$\mathbf{1}_H \leq \eta_\kappa * \mathbf{1}_{H_1}.$$

We also have that for $z \in \mathbb{C}$ and $\kappa < \kappa_0$

$$|(\eta_\kappa * \varphi)(z) - \varphi(z)| \leq \int_{-\kappa}^{\kappa} \eta_\kappa(t) |\varphi(g_{-t}z) - \varphi(z)| \, dt < \epsilon.$$

A sequence of scalings. Since $\mathbf{1}_H \in B_c(\mathbb{C})$, we have by [10], $\widehat{\mathbf{1}_H} \in L^2(\mu)$ so

$$\int_{\mathcal{H}} \widehat{\mathbf{1}_H} \, d\rho_{\mathcal{H}} = c_{\mathcal{H}} m(H) < \infty.$$

Let $\epsilon > 0$, and fix a sequence of scalings $\{\kappa_n\}$ with $\kappa_n \to 0$ and choose N so that for $n \geq N$ and $z \in H$,

(6.5.3) $\quad |f_n(z) - \varphi(z)| < \epsilon \quad$ and $\quad |(\eta_{\kappa_n} * \varphi)(z) - \varphi(z)| < \epsilon.$

Summing over holonomies. Since the supports of f_n and φ are subsets of H, we can sum over $\Lambda_{\omega'}$ for any $\omega' \in \mathcal{H}$ to get

(6.5.4) $\quad |\widehat{f_n}(\omega') - \widehat{\varphi}(\omega')| < \epsilon \widehat{\mathbf{1}_H}(\omega') \quad$ and $\quad |(\eta_{\kappa_n} * \varphi)^{SV}(\omega') - \widehat{\varphi}(\omega')| < \epsilon \widehat{\mathbf{1}_H}(\omega').$

Fix $n \geq N$. Using that $\mathbf{1}_{H_1}$ is K-finite, we can apply Black Box 6.5.2 to obtain, for almost every $\omega \in \mathcal{H}$, a $T = T(\omega, \kappa_0)$ large enough so that for all $t \geq T$,

(6.5.5) $$|A_t(\eta_{\kappa_n} * \mathbf{1}_{H_1})^{SV}(\omega) - m(H_1)| < \epsilon$$

and

(6.5.6) $$A_t(\mathbf{1}_H)^{SV}(\omega) \leq A_t(\eta_{\kappa_n} * \mathbf{1}_{H_1})^{SV}(\omega) \leq m(H_1) + \epsilon.$$

Taking limits. Fix ω from our full measure set and $t \geq T$. Applying the left-hand inequality in (6.5.4) and combining with (6.5.6) yields

(6.5.7) $$|A_t(\eta_{\kappa_n} * f_n)^{SV}(\omega) - A_t(\eta_{\kappa_n} * \varphi)^{SV}(\omega)| \leq \epsilon A_t(\eta_{\kappa_n} * \mathbf{1}_{H_1})^{SV}(\omega)$$
$$< \epsilon[\epsilon + m(H_1)].$$

Using the right-hand inequality in (6.5.4) and combining with (6.5.6) yields

(6.5.8) $$|A_t(\eta_{\kappa_n} * \varphi)^{SV}(\omega) - A_t \varphi^{SV}(\omega)| \leq A_t|(\eta_{\kappa_n} * \varphi)^{SV}(\omega) - \varphi^{SV}(\omega)|$$
$$\leq \epsilon A_t(\mathbf{1}_H)^{SV}(\omega)$$
$$\leq \epsilon[\epsilon + m(H_1)].$$

6.5. Equidistribution of circles

Applying Nevo's theorem. By Black Box 6.5.2, for each f_n and almost every ω,

$$\lim_{\tau \to \infty} A_\tau(\widehat{\eta_{\kappa_n} * f_n})(\omega) = \lim_{\tau \to \infty} \int_{-\infty}^{\infty} \eta_{\kappa_n}(t) \left(A_{\tau-t}\widehat{f_n}\right)(\omega) dt$$

$$= \lim_{\tau \to \infty} \int_{-\infty}^{\infty} \eta_{\kappa_n}(\tau - s) \left(A_s\widehat{f_n}\right)(\omega) ds$$

$$= \int_{\mathcal{H}} \widehat{f_n} \, d\rho_{\mathcal{H}} = c_{\mathcal{H}} \int_{\mathbb{C}} f_n \, dm,$$

where m is Lebesgue measure on $\mathbb{C}$.

Thus for a.e. ω we can choose $T = T(\omega)$ large enough so that for all $t \geq T$,

(6.5.9) $$\left| A_t(\eta_{\kappa_n} * f_n)^{\text{SV}}(\omega) - c_{\mathcal{H}} \int_{\mathbb{C}} f_n \, dm \right| < \epsilon.$$

Finally, we can use (6.5.3) to say that since f_n and φ have support in H_1,

(6.5.10) $$\left| \int_{\mathbb{C}} f_n \, dm - \int_{\mathbb{C}} \varphi \, dm \right| \leq \epsilon \, m(H_1).$$

Putting it all together. Now applying the triangle inequality together with (6.5.7), (6.5.8), (6.5.9), and (6.5.10) we conclude that for almost every $\omega \in \mathcal{H}$ and each n, there is $T = T(\varphi, \omega)$ so that for all $t \geq T$,

(6.5.11) $$\left| A_t\widehat{\varphi}(\omega) - c_{\mathcal{H}} \int_{\mathbb{C}} \varphi \, dm \right| \leq 2\epsilon[\epsilon + m(H_1)] + \epsilon + \epsilon m(H_1).$$

Conclusion. Since $m(H)$ and $m(H_1)$ are fixed constants and ϵ is arbitrary, we conclude for almost every ω,

(6.5.12) $$\lim_{t \to \infty} A_t\widehat{\varphi}(\omega) = c_{\mathcal{H}} \int_{\mathbb{C}} \varphi \, dm.$$

We have thus proved Theorem 6.5.1. □

Back to counting. To finish the proof of Theorem 6.2.1, we combine Theorem 6.5.1 with the triangle trick from §6.4.2. However, $\mathbf{1}_{\mathcal{T}}$ is not a continuous function (though of course it is bounded and of compact support). The following Black Box (a simpler version of [12, Lemma 3.4]), whose proof we leave as a nontrivial exercise, allows us to approximate $\mathbf{1}_{\mathcal{T}}$ by continuous compactly supported functions f_ϵ in such a way that the limits of the averaging operators A_t are close.

Black Box 6.5.4. *Let $\mathcal{T}$ be the triangle as defined in §6.4.2. Then for all ϵ there exists a function $f_\epsilon \in C_c(\mathbb{C})$ and $T \geq 0$ so that for all $t \geq T$ and μ-a.e. ω,*

$$(6.5.13) \qquad \left|A_t\left(f_\epsilon - \mathbf{1}_{\mathcal{T}}\right)^{SV}(\omega)\right| < \epsilon \quad \text{and} \quad \left|\int_{\mathcal{H}} (f_\epsilon - \mathbf{1}_{\mathcal{T}})^{SV} d\rho_{\mathcal{H}}\right| < \epsilon.$$

Using this, we obtain our almost sure counting result Theorem 6.2.1. □

6.6. Further counting results

Further counting results. The connection between the dynamics of the $SL(2, \mathbb{R})$-action and counting can be used to prove counting results beyond Theorem 6.2.1.

Effective results and exponential mixing. As we saw above, the proof of Theorem 6.2.1 relied on Nevo's equidistribution theorem Black Box 6.5.2 for the action of $SL(2, \mathbb{R})$ on $\mathcal{H}$, which uses in a crucial way the Howe-Moore theorem, Black Box 5.3.2, which guarantees mixing of the action. This equidistribution theorem and Theorem 6.2.1 were subsequently made *effective* by work of Nevo-Rühr-Weiss [136], who, using the Avila-Gouëzel-Yoccoz [22] exponential mixing result Theorem 5.3.6, showed the existence of an $\alpha < 2$ such that

$$\left|N(\omega, R) - c_{\mathcal{H}} \pi R^2\right| = o(R^\alpha).$$

Averaged quadratic asymptotics and orbit closures. Using the orbit closure and equidistribution results of Eskin-Mirzakhani [61], Thereom 6.2.1 was strengthened by Eskin-Mirzakhani-Mohammadi [62] to obtain *averaged quadratic asymptotics* for every ω, using the triangle trick and the Cesáro convergence of the averages $A_t \widehat{\varphi}$ as $t \to \infty$. Precisely, they showed:

Theorem 6.6.1. *For every translation surface ω, there is a $0 < c_\omega < \infty$ such that*

$$(6.6.1) \qquad \lim_{T \to \infty} \frac{1}{T} \int_0^T N(\omega, e^t) e^{-2t} dt = c_\omega.$$

Other measures. The arguments of Theorem 6.2.1 can be used to show almost sure counting for any ergodic $SL(2, \mathbb{R})$-invariant measure ν on $\mathcal{H}$, with the constant $c_{\mathcal{H}}(\nu)$ once again being the constant that would appear in the analog of the Siegel-Veech formula for ν.

The constants. We note that Theorem 6.2.1 implies that for $\mu_{\mathcal{H}}$-almost every ω, the constant $c_\omega = c_{\mathcal{H}}$. It is still an open question as to whether (6.6.1) can be strengthened to the statement that for all $\omega \in \mathcal{H}$ there is a c_ω such that

$$\lim_{R \to \infty} \frac{N(\omega, R)}{\pi R^2} = c_\omega.$$

Counting pairs. Using similar ideas as in the proof of Theorem 6.2.1 and the L^2-results on the Siegel-Veech transform of Athreya-Cheung-Masur [10], Athreya-Fairchild-Masur [12] showed how to obtain almost sure (with respect to $\mu_\mathcal{H}$) quadratic asymptotics for counting *pairs* of saddle connections whose holonomy vectors span a parallelogram of bounded area. These results were subsequently effectivized (as in Nevo-Rühr-Weiss [136]) by Bonnafoux [27, Theorem 1.2].

6.7. Counting and volumes

Degenerations and boundary strata. Just as in the setting of the space of lattices X_n, there is a close connection between the intrinsic volumes $\mu_\mathcal{H}(\mathcal{H})$ and the counting constants $c_\mathcal{H}$, obtained by trying to understand the behavior of the Siegel-Veech transform applied to the indicator function $\mathbf{1}_{B(0,\epsilon)}$ of the ball $B(0,\epsilon)$ of radius ϵ in $\mathbb{C}$ and neighborhoods of ∞ in $\mathcal{H}$. By abuse of notation, we write $\mathbf{1}_\epsilon$ for $\mathbf{1}_{B(0,\epsilon)}$. Following Eskin-Masur-Zorich [58, §3.3], we define $\mathcal{H}^\epsilon \subset \mathcal{H}$ to be the set of translation surfaces with an ϵ-short saddle connection; that is,

$$\mathcal{H}^\epsilon = \{\omega \in \mathcal{H} : \Lambda_\omega \cap B(0,\epsilon) \neq \emptyset\}.$$

We split $\mathcal{H}^\epsilon$ further into two pieces,

$$\mathcal{H}^{\epsilon,\text{thick}} = \{\omega \in \mathcal{H} : \#(\Lambda_\omega \cap B(0,\epsilon)) = 1\},$$

consisting of those surfaces with a saddle connection of length at most ϵ and such that any other saddle connection of length at most ϵ is homologous to it, and $\mathcal{H}^{\epsilon,\text{thin}} = \mathcal{H}^\epsilon \setminus \mathcal{H}^{\epsilon,\text{thick}}$, which consists of surfaces with multiple nonhomologous saddle connections of length at most ϵ. Since $\hat{\mathbf{1}}_\epsilon$ is supported on $\mathcal{H}^\epsilon$ and is identically 1 on $\mathcal{H}^{\epsilon,\text{thick}}$, we can use the Siegel-Veech formula Theorem 6.3.1 to write

$$(6.7.1) \quad c_\mathcal{H} \pi \epsilon^2 = c_\mathcal{H} \int_\mathbb{C} \mathbf{1}_\epsilon dm = \int_\mathcal{H} \hat{\mathbf{1}}_\epsilon d\rho_\mathcal{H}$$

$$= \frac{1}{\mu_\mathcal{H}(\mathcal{H})} \int_\mathcal{H} \hat{\mathbf{1}}_\epsilon d\mu_\mathcal{H} = \frac{1}{\mu_\mathcal{H}(\mathcal{H})} \int_{\mathcal{H}^\epsilon} \hat{\mathbf{1}}_\epsilon d\mu_\mathcal{H}$$

$$= \frac{1}{\mu_\mathcal{H}(\mathcal{H})} \left(\int_{\mathcal{H}^{\epsilon,\text{thick}}} \hat{\mathbf{1}}_\epsilon d\mu_\mathcal{H} + \int_{\mathcal{H}^{\epsilon,\text{thin}}} \hat{\mathbf{1}}_\epsilon d\mu_\mathcal{H} \right)$$

$$= \frac{1}{\mu_\mathcal{H}(\mathcal{H})} \left(\mu_\mathcal{H}(\mathcal{H}^{\epsilon,\text{thick}}) + \int_{\mathcal{H}^{\epsilon,\text{thin}}} \hat{\mathbf{1}}_\epsilon d\mu_\mathcal{H} \right).$$

The thin part. While the function $\hat{1}_\epsilon$ is unbounded on the thin part $\mathcal{H}^{\epsilon,\text{thin}}$, using ideas from Eskin-Masur [57], Eskin-Masur-Zorich [58, Proposition 3.3] showed the following bound:

Black Box 6.7.1. *The integral of $\hat{1}_\epsilon$ on the thin part $\mathcal{H}^{\epsilon,\text{thin}}$ is negligible; that is, as $\epsilon \to 0$,*

$$\int_{\mathcal{H}^{\epsilon,\text{thin}}} \hat{1}_\epsilon d\mu_{\mathcal{H}} = o(\epsilon^2). \tag{6.7.2}$$

Combining Black Box 6.7.1 with (6.7.1), we obtain

$$\lim_{\epsilon \to 0} \frac{1}{\pi\epsilon^2} \frac{\mu_{\mathcal{H}}(\mathcal{H}^\epsilon)}{\mu_{\mathcal{H}}(\mathcal{H})} = c_{\mathcal{H}}. \tag{6.7.3}$$

Configurations. In fact, [58] showed this for any *configuration* of saddle connections. A configuration $\mathcal{C}$ is any $SL(2,\mathbb{R})$-equivariant assignment

$$\omega \longmapsto \Lambda^{\mathcal{C}}_\omega \subset \Lambda_\omega,$$

for example, the set of holonomy vectors of saddle connections connecting two zeros of a particular order. For any configuration, there is an analog of the Siegel-Veech formula Theorem 6.3.1 with a constant $c_{\mathcal{C},\mathcal{H}}$. Given a configuration $\mathcal{C}$, they further showed that if $\mathcal{H}_{\mathcal{C}}$ is the stratum obtained by degenerating a surface in $\mathcal{H}$ along a saddle connection in $\mathcal{C}$, there is a combinatorial constant $M = M_{\mathcal{C},\mathcal{H}}$ so that

$$c_{\mathcal{C},\mathcal{H}} = M \frac{\mu_{\mathcal{H}_{\mathcal{C}}}(\mathcal{H}_{\mathcal{C}})}{\mu_{\mathcal{H}}(\mathcal{H})}.$$

Computing volumes. To use these ideas to explicitly compute the volumes, one needs to compute the constants M explicitly, and to be able to compute the constants $c_{\mathcal{H}}$ or, vice versa, to compute the constants $c_{\mathcal{H}}$, one needs explicit information about the volumes $\mu_{\mathcal{H}}(\mathcal{H})$.

Genus 2. As an important example, in genus 2 we have two strata, $\mathcal{H}(1,1)$ and $\mathcal{H}(2)$. If we consider $\mathcal{H} = \mathcal{H}(1,1)$ and our configuration $\mathcal{C}$ as consisting of saddle connections connecting the two distinct zeros, a special case of [58, Formula 8.2] states that

$$M_{\mathcal{C},\mathcal{H}} = 3.$$

The volumes [58, Table 1] in genus 2 were shown to be

$$\mu_{\mathcal{H}(1,1)}(\mathcal{H}(1,1)) = \frac{\pi^4}{135}, \quad \mu_{\mathcal{H}(2)}(\mathcal{H}(2)) = \frac{\pi^4}{120}.$$

The stratum $\mathcal{H}(2)$ is the degeneration of $\mathcal{H}(1,1)$ by degenerating along the saddle connection joining distinct zeros. We get

$$c_{\mathcal{C},\mathcal{H}} = 3\frac{135}{120} = \frac{27}{8}.$$

6.7. Counting and volumes

Other methods. We will discuss below in §7.8 methods for computing $\mu_{\mathcal{H}}(\mathcal{H})$ related to counting *integer points* in strata Ω, and there are remarkable connections, first conjectured by Kontsevich-Zorich [**108, 109**] between the constants $c_{\mathcal{H}}$ and the Lyapunov exponents for the Teichmüller geodesic flow we discussed in §5.4, which have been explored by, for example, Eskin-Kontsevich-Zorich [**54**], Athreya-Eskin-Zorich [**11**], Chen-Möller-Zagier [**35**], and Goujard [**80**], among others.

Boundaries. We mention that there has been important recent work of Bainbridge-Chen-Gendron-Grushevsky-Möller [**24**] on constructing boundaries of strata (and other moduli spaces of differentials) in ways that are well adapted to compactifications of moduli space and the $GL^+(2, \mathbb{R})$-action.

Chapter 7

Lattice Surfaces

We now turn to a special family of surfaces, those with a large group of affine symmetries, known as *lattice surfaces* or *Veech surfaces*. These surfaces have played a very important role in the historical development of the theory of translation surfaces. Although the set of lattice surfaces forms a set of measure 0 in any stratum, many of the basic examples of translation surfaces fall into this family and were among the first translation surfaces studied, with respect to both dynamical and counting properties. They form a *dense* set in any stratum. We first define the notions of affine diffeomorphisms and Veech groups of translation surfaces in §7.1 and then review a series of important examples in §7.2. We discuss some important connections between the flat geometry of the surface and the Veech group in §7.3. We state and prove an important theorem of Veech [**164**] on the *optimal dynamics* of linear flows on lattice surfaces in §7.4. In §7.5 we show how classical results on dynamics on hyperbolic surfaces can be used to implement the counting strategies from Chapter 6 in a simpler way for lattice surfaces. Lattice surfaces have many different equivalent characterizations, which we review in §7.6. In §7.7 we give various characterizations of lattice surfaces that can be built out of other lattice surfaces via branched coverings, and perhaps the most important example of surfaces that arise in this way are *square-tiled surfaces* or *origami*, branched covers of flat tori, which help link the study of translation surfaces to number theory in deep and interesting ways. We discuss these surfaces in §7.8. This chapter, which is lighter on details than the previous chapters, uses in a crucial way the surveys of Hubert-Schmidt [**92**] and Massart [**115**] and the more recent exposition of McMullen [**128**], which focuses on the classification problem for lattice surfaces.

7.1. Affine diffeomorphisms and Veech groups

Let $\omega \in \mathcal{H}$ be a translation surface. An orientation-preserving diffeomorphism φ of the underlying surface X is said to be *affine* if φ is an affine transformation in the local flat coordinates defined by ω and $\varphi_*\omega = \omega$. Since the transition maps in these coordinates are translations, the linear part (that is, the derivative) $A = D\varphi$ of φ is constant.

Definition 7.1.1. Let $\omega \in \mathcal{H}$. We define the affine diffeomorphism group

$$\mathrm{Aff}^+(\omega) := \{\varphi : \varphi \text{ is affine and orientation preserving}\}.$$

We define the *Veech group*

$$\Gamma(\omega) := D(\mathrm{Aff}^+(\omega)) \subset SL(2, \mathbb{R})$$

of ω to be the image of $\mathrm{Aff}^+(\omega)$ under the derivative map D.

Stabilizers. The group $\Gamma(\omega)$ is the stabilizer of ω under the $SL(2, \mathbb{R})$-action on $\mathcal{H}$. The ergodicity of the $SL(2, \mathbb{R})$-action with respect to $\mu_\mathcal{H}$ implies that for $\mu_\mathcal{H}$-almost every ω, $\Gamma(\omega)$ is trivial. We leave this proof (and the closely related proof that $\mathrm{Aff}^+(\omega)$ is almost surely trivial) as an exercise. As a note, for hyperelliptic translation surfaces ω, the involution ι sends ω to $-\omega$ (see, for example, [107, §2.1, page 638]) so we do not view it as an element of $\mathrm{Aff}^+(\omega)$.

Exercise 7.1. *Assume $g \geq 2$. Prove that for $\mu_\mathcal{H}$-almost every $\omega \in \mathcal{H}$, $\mathrm{Aff}^+(\omega)$ and $\Gamma(\omega)$ are trivial.*

Conjugates. Since $\Gamma(\omega)$ is the stabilizer of ω under the $SL(2, \mathbb{R})$-action, we have

$$\Gamma(g \cdot \omega) = g\Gamma(\omega)g^{-1}.$$

Kernels. When the group $\Gamma(\omega)$ is nontrivial (and we will see many examples below), an important observation of Veech [164] is that the elements of the kernel $\ker(D, \omega)$ of the map

$$D : \mathrm{Aff}^+(\omega) \to \Gamma(\omega)$$

are automorphisms of the underlying Riemann surface X. For $g \geq 2$, we have the following Black Box due to Hurwitz [94]; see, for example, Farb-Margalit [64, Theorem 7.4]:

Black Box 7.1.2 ([64, Theorem 7.4]). *If X is a compact Riemann surface of genus $g \geq 2$, the automorphism group $\mathrm{Aut}^+(X)$ is finite and has cardinality at most $84(g-1)$.*

7.1. Affine diffeomorphisms and Veech groups

7.1.1. Discreteness. We describe an argument due to Vorobets [168] that shows $\Gamma(\omega)$ is a *Fuchsian group*, that is, a discrete subgroup of $SL(2,\mathbb{R})$. As with many facts about $\Gamma(\omega)$, this result is originally due to Veech [164].

Lemma 7.1.3. *Let $\omega \in \mathcal{H}$. Then $\Gamma(\omega)$ is discrete.*

Proof. We recall from Lemma 2.4.7 that the set of holonomy vectors Λ_ω is a discrete subset of $\mathbb{C}$ and that for $g \in SL(2,\mathbb{R})$,

$$\Lambda_{g\cdot\omega} = g \cdot \Lambda_\omega.$$

Suppose now that $\{\gamma_n\} \subset \Gamma(\omega)$ is a sequence of elements in $\Gamma(\omega)$ with

$$\gamma_n \longrightarrow I,$$

where I is the identity matrix. Take any $z, w \in \Lambda_\omega$ linearly independent over $\mathbb{R}$ (that is, $z/w \notin \mathbb{R}$). We have

$$\gamma_n z \longrightarrow z, \quad \gamma_n w \longrightarrow w.$$

Since Λ_ω is discrete, there is an $N > 0$ so that for $n \geq N$,

$$\gamma_n z = z, \quad \gamma_n w = w,$$

and therefore $\gamma_n = I$ for $n \geq N$. □

7.1.2. Noncompactness. The groups that can arise as $\Gamma(\omega)$ are further restricted beyond discreteness—they can never be cocompact.

Lemma 7.1.4. *Let $\omega \in \mathcal{H}$. Then $\Gamma(\omega)$ is not a cocompact subgroup of $SL(2,\mathbb{R})$; that is, the quotient $SL(2,\mathbb{R})/\Gamma(\omega)$ is noncompact.*

Proof. This proof is very similar to that of Proposition 3.5.5, which shows that strata $\mathcal{H}$ are noncompact. We will define a continuous function on $SL(2,\mathbb{R})/\Gamma(\omega)$ which does not achieve its minimum. Recall the systole function ℓ from Definition 3.5.3. For $g \in SL(2,R)$, define the continuous positive function f_ω by

$$f_\omega(g) = \ell(g \cdot \omega) = \min_{z \in g \cdot \Lambda_\omega} |z|.$$

Since $\Gamma(\omega)$ stabilizes Λ_ω, f_ω descends to a function on $SL(2,\mathbb{R})/\Gamma$. Take any $z = re^{i\theta} \in \Lambda_\omega$. Then

$$f_\omega(g_t r_{\pi/2-\theta}) \leq |g_t r_{\pi/2-\theta} z| = e^{-t/2}|z|,$$

so for any $\epsilon > 0$, there is a $g \in SL(2,\mathbb{R})$ so that $f_\omega(g) < \epsilon$. Thus f_ω does not achieve its minimum on $SL(2,\mathbb{R})/\Gamma(\omega)$, so $\Gamma(\omega)$ is not cocompact. □

Infinitely generated examples. We note that Veech groups $\Gamma(\omega)$ can in fact be extremely complicated—for example, they do not have to be finitely generated. Hubert-Schmidt [93] gave concrete examples of surfaces with *infinitely generated* Veech groups. One such example is a double cover of the regular octagon, branched over the zero and one other point. These examples have quite interesting ergodic properties, as we discuss in §7.6.2 below.

7.1.3. Lattice surfaces. We now state the definition of a *lattice surface*.

Definition 7.1.5. We say $\omega \in \mathcal{H}$ is a *lattice surface* if $\Gamma(\omega)$ is a lattice in $SL(2,\mathbb{R})$; that is, the Haar measure $\mu(SL(2,\mathbb{R})/\Gamma(\omega)) < \infty$. In particular they must be finitely generated.

Parabolic elements. The non-cocompactness of $\Gamma(\omega)$ implies that if ω is a lattice surface, then $\Gamma(\omega)$ will contain *parabolic* elements of $SL(2,\mathbb{R})$. Conjugacy classes of maximal parabolic subgroups correspond to *cusps* (finite area noncompact regions of $SL(2,\mathbb{R})/\Gamma$) (see, for example, [40, §I.4]). We recall that $g \in SL(2,\mathbb{R})$ is *parabolic* (respectively, *elliptic* or *hyperbolic*) if $|\operatorname{tr}(g)| = 2$ (respectively, $|\operatorname{tr}(g)| < 2$ or $|\operatorname{tr}(g)| > 2$). Any parabolic element of $SL(2,\mathbb{R})$ is conjugate to a matrix of the form

$$h_s = \begin{pmatrix} 1 & s \\ 0 & 1 \end{pmatrix}$$

for some $s \in \mathbb{R}$, and such an element in $\Gamma(\omega)$ corresponds to a shear of the surface. We will see below in §7.3 how to build these types of elements out of *cylinder decompositions* of the surface ω where the cylinders have rationally related moduli.

7.2. Examples

We discuss several examples of families of lattice surfaces. We will not give detailed proofs in this section that these are lattice surfaces or of computations of the Veech groups $\Gamma(\omega)$, instead postponing these discussions to §7.6 and giving references where appropriate.

7.2.1. The torus and its covers. We start with our motivating example of a translation surface, the square torus $(\mathbb{C}/\mathbb{Z}[i], dz)$. We recall from Exercise 1.1 that the stabilizer of this surface under the $SL(2,\mathbb{R})$-action is $SL(2,\mathbb{Z})$. We say a translation surface (X, ω) is *square-tiled* or an *origami* if it is a cover of $(\mathbb{C}/\mathbb{Z}[i], dz)$ branched over 0. That is, there is a covering map

$$p : X \to \mathbb{C}/\mathbb{Z}[i],$$

branched over 0, with $\omega = p^*(dz)$. The zeros of ω are contained in the preimage $p^{-1}(0)$. If the degree of p is n, the surface is obtained from a polygon which is tiled by n squares.

7.2. Examples

Areas. If the underlying torus has area 1, the resulting surface has area n, so to obtain a unit-area surface, we normalize the torus to have area n^{-1}; that is, the side lengths of the squares will be $n^{-1/2}$.

Commensurability. Gutkin-Judge [81] showed that for any square-tiled ω, the Veech group $\Gamma(\omega)$ is *commensurable* to $SL(2,\mathbb{Z})$; that is, there is a common finite index subgroup Γ' of Γ and $SL(2,\mathbb{Z})$. Indeed, $\Gamma(\omega)$ is commensurable to $SL(2,\mathbb{Z})$ if and only if ω is a torus cover (so a translation surface tiled by parallelograms). We will discuss this family of examples in much more detail in §7.8, in particular discussing combinatorial constructions and the fact that square-tiled surfaces provide the natural notion of *integer points* in strata Ω.

A 3-square L. A first concrete example of a square-tiled surface is the surface ω_{3L} constructed out of an L-shape made with 3 squares, illustrated below in Figure 7.1. This surface, in $\Omega(2)$, is a degree 3 cover of the torus. The corners of the squares of the L are preimages of 0, and they collapse to one point with angle 6π on the surface. We leave as an exercise:

Exercise 7.2. *Show that the Veech group $\Gamma(\omega_{3L})$ of the 3-square L ω_{3L} contains*

$$\begin{pmatrix} 1 & 2 \\ 0 & 1 \end{pmatrix} \quad \text{and} \quad \begin{pmatrix} 1 & 0 \\ 2 & 1 \end{pmatrix}$$

but not

$$\begin{pmatrix} 1 & 1 \\ 0 & 1 \end{pmatrix} \quad \text{and} \quad \begin{pmatrix} 1 & 0 \\ 1 & 1 \end{pmatrix}.$$

Hint: Draw the pictures of the sheared squares, and try to cut and reassemble into the original square.

Figure 7.1. A 3-square L translation surface $\omega_{3L} \in \Omega(2)$.

7.2.2. L-shapes. We have seen that L-shapes lead to a *family* of translation surfaces, by varying the lengths of the sides. We see a 2-parameter family, first explored by McMullen [126], illustrated in Figure 7.2.

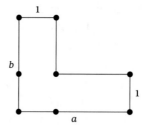

Figure 7.2. An element $\omega_{L(a,b)}$ of a 2-parameter family in $\Omega(2)$ [**128**, Figure 2.1].

Lattice L-shapes. For particular values of a and b, the surface $\omega_{L(a,b)}$ is a lattice surface. We will discuss seminal results of McMullen [**123**] and Calta [**31**] which determine explicitly these values in §7.7. For now we focus on a few important examples.

The golden L. An important L-shaped table is the *golden L*, $\omega_{L(\phi,\phi)}$, which we denote ω_ϕ. Here $\phi = \frac{1+\sqrt{5}}{2}$ is the golden ratio. It is a nontrivial fact that

$$\Gamma(\omega_\phi) = \left\langle \begin{pmatrix} 0 & 1 \\ -1 & 0 \end{pmatrix}, \begin{pmatrix} 1 & \phi \\ 0 & 1 \end{pmatrix} \right\rangle.$$

We leave as a nontrivial exercise that these elements are indeed in the Veech group of ω_ϕ.

Exercise 7.3. *Show that*

$$\begin{pmatrix} 1 & \phi \\ 0 & 1 \end{pmatrix} \quad \text{and} \quad \begin{pmatrix} 0 & 1 \\ -1 & 0 \end{pmatrix}$$

are elements of $\Gamma(\omega_\phi)$.

Hecke triangle groups. Both the torus and the golden L are examples of translation surfaces where the Veech group is an element of a family of very classical groups called *Hecke triangle groups*. They are discrete subgroups of $SL(2, \mathbb{R})$ associated to reflections in sides of hyperbolic triangles. Given $k_i \in \mathbb{N} \cup \{\infty\}$ with

$$\frac{1}{k_1} + \frac{1}{k_2} + \frac{1}{k_3} < 1$$

we define the *Hecke triangle group* $\Delta(k_1, k_2, k_3)$ to be the index 2 subgroup of orientation-preserving isometries of the hyperbolic plane $\mathbb{H}^2$ inside the group generated by reflections in sides of a hyperbolic triangle in $\mathbb{H}^2$ with angles $\pi/k_1, \pi/k_2, \pi/k_3$. We note that the k_i are allowed to be ∞, in which case the corresponding vertex of the triangle is on the boundary at infinity of the hyperbolic plane and the vertex angle is 0. See, for example, Schmidt-Sheingorn [**148**] for a more general discussion of Hecke triangle groups and their generators.

7.2. Examples

7.2.3. Regular and semiregular polygons. Some of the most natural examples of translation surfaces arise from regular and semiregular polygons. We discussed in §2.1.1 many examples of regular polygons which lead to translation surfaces, and we revisit these discussions in the context of lattice surfaces.

The regular octagon. Recall from §2.1.1 that identifying parallel sides of a unit-area regular octagon leads to a surface in $\mathcal{H}(2)$, which we denote ω_8. A direct computation shows (see, for example, Smillie-Ulcigrai [154]) that $\Gamma(\omega_8)$ is conjugate to the Hecke $\Delta(4, \infty, \infty)$ group (in particular, it has two distinct cusps), and

$$\Gamma(\omega_8) = \left\langle \begin{pmatrix} 1/\sqrt{2} & -1/\sqrt{2} \\ 1/\sqrt{2} & 1/\sqrt{2} \end{pmatrix}, \begin{pmatrix} 1 & 2(1+\sqrt{2}) \\ 0 & 1 \end{pmatrix} \right\rangle.$$

The octagonal L. In fact, the surface ω_8 is in the same $GL^+(2, \mathbb{R})$-orbit as an L-shaped table. We leave this as a nontrivial exercise (see Uyanik-Work [160, §2] for a hint).

Exercise 7.4. *Show that applying the matrix*

$$\begin{pmatrix} 1 & -(1+\sqrt{2}) \\ 0 & 1 \end{pmatrix}$$

and an appropriate real scaling transforms $\omega_8 \in \mathcal{H}(2)$ into the L-shaped table $\omega_{L(\sqrt{2},\sqrt{2}+1)} \in \Omega(2)$.

Regular evengons. More generally, Veech [164] showed that the translation surfaces $\omega_{4g} \in \mathcal{H}(2g-2)$ and $\omega_{4g+2} \in \mathcal{H}(g-1, g-1)$ obtained by identifying the opposite sides of the unit-area regular $4g$ or $4g+2$-gon lead to a lattice surface and explicitly computed the Veech groups $\Gamma(\omega_{4g})$ and $\Gamma(\omega_{4g+2})$. Recall that we have

$$r_{\pi/n} = \begin{pmatrix} \cos \pi/n & -\sin \pi/n \\ \sin \pi/n & \cos \pi/n \end{pmatrix}, \quad h_{\cot(\pi/n)} = \begin{pmatrix} 1 & 2\cot(\pi/n) \\ 0 & 1 \end{pmatrix}.$$

We have the following:

Black Box 7.2.1 ([164, Theorem 5.8]). *For $k \geq 4$,*

$$\Gamma(\omega_{2k}) = \langle r_{\pi/2k}^2, h_{\cot(\pi/2k)}, r_{\pi/2k} h_{\cot(\pi/2k)} r_{\pi/2k}^{-1} \rangle,$$

and it is isomorphic to the Hecke triangle group $\Delta(k, \infty, \infty)$ and has two distinct cusps.

Double regular oddgons. Starting with a regular $2k+1$-gon of area $1/2$, we can double it along a side to obtain a unit-area topological $4k$-gon. Identifying parallel sides, we obtain a surface $\omega_{2k+1} \in \mathcal{H}(2k-2)$. We have

Black Box 7.2.2 ([164], Theorem 5.8]). *For $k \geq 1$,*
$$\Gamma(\omega_{2k+1}) = \langle r_{\pi/2k}, h_{\cot(\pi/2k)} \rangle,$$
and it is isomorphic to the Hecke triangle group $\Delta(2, k, \infty)$ and has one cusp.

Tori. We leave as an exercise:

Exercise 7.5. *Check that ω_3 is a rhombic flat torus. Show that $\Gamma(\omega_3)$ is conjugate to $SL(2, \mathbb{Z})$ and is isomorphic to $\Delta(2, 3, \infty)$.*

The double pentagon and the golden L. An important example of a double regular oddgon is the case $k = 2$, of the double pentagon, illustrated in Figure 7.3. We leave as an exercise that this surface is in the same $GL^+(2, \mathbb{R})$-orbit as the golden L.

Exercise 7.6. *Show there is a $g \in GL^+(2, \mathbb{R})$ so that*
$$g \cdot \omega_5 = \omega_\phi.$$

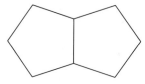

Figure 7.3. The double pentagon. Parallel sides are identified by translation to form a translation surface in $\Omega(2)$.

Semiregular polygons. More generally, there are explicit constructions of lattice surfaces due to Ward [169] (motivated by the study of triangular billiards), Bouw-Möller [28] (using algebraic ideas), and Hooper [85] (using explicit flat constructions from semiregular polygons). Wright [172] showed that these constructions in fact coincide.

7.2.4. Platonic covers. The translation surfaces which arise as covers of the k-differentials on the sphere induced by the surfaces of Platonic solids, which we discussed in §2.6.5, are lattice surfaces, as discussed in detail in Athreya-Aulicino-Hooper [6]. The double cover of the tetrahedron is a flat (rhombic) torus, the degree 4 cover of the cube is a genus 9 surface square-tiled by 24 squares, and the covers of the octahedron (degree 3, genus 4) and the icosahedron (degree 6, genus 25) are tiled by rhombi, and thus by our discussion in §7.2.1 all have Veech groups commensurable to $SL(2, \mathbb{Z})$. The degree 10, genus 81 translation cover of the dodecahedron is tiled by 120 pentagons and in fact

7.2. Examples

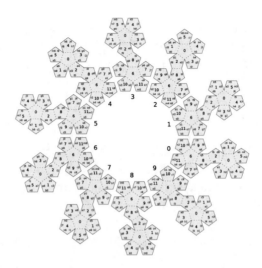

Figure 7.4. The genus 81 translation surface arising from the 10-cover of the dodecahedron [**6**, Figure 1]. Reprinted by permission of the publisher (Taylor & Francis Ltd, http://www.tandfonline.com).

is a degree 60 cover of the double pentagon, branched over the singular point; see Figure 7.4, taken from [**6**].

7.2.5. Branched covers. More generally, following [**123**], we say a translation surface ω (with underlying Riemann surface X) is the *pullback* of another η (with underlying Riemann surface Y) if there is a holomorphic map $f : X \to Y$ of the underlying Riemann surfaces such that $\omega = f^*(\eta)$. Note that if Σ_X and Σ_Y are the sets of zeros of ω and η, respectively, $f^{-1}(\Sigma_Y) \subset \Sigma_X$. If in addition $f(\Sigma_X) \subset \Sigma_Y$, we will say ω is a *translation covering* of η. We already implicitly introduced these notions in §7.2.1 (and we discuss them further in the context of primitive surfaces in §7.7). We now state as a Black Box the general result of Gutkin-Judge [**81**] which relates the Veech groups $\Gamma(\omega)$ and $\Gamma(\omega')$.

Black Box 7.2.3 ([**81**, Theorem 4.9]). *If ω is a translation cover of η, then the Veech groups $\Gamma(\omega)$ and $\Gamma(\eta)$ are commensurable. In particular, ω is a lattice surface if and only if η is.*

7.2.6. Arnoux-Yoccoz surfaces. We discussed in §7.1.1 above examples of translation surfaces with *infinitely generated* Veech groups, but there are many other behaviors which can occur. An important series of examples of translation surfaces with finitely generated Veech groups which are *not* lattices but have (large) finitely generated Veech groups but *no parabolic elements* were introduced by Arnoux-Yoccoz [**4**]. They built translation surfaces in each genus $g \geq 3$, built out of an interval exchange map whose length parameters are determined by a (positive, real) root α of the polynomial $x^g + x^{g-1} + \cdots + x - 1$. Hubert-Lanneau-Möller [**91**] showed that these surfaces in fact have distinct (nonconjugate) hyperbolic elements in their stabilizers. Bowman [29] gave an alternate geometric description and in fact constructed an interesting limit (as $g \to \infty$) for these surfaces, which is an *infinite-type* translation surface.

7.3. Cusps, cylinders, and shears

We record in this section an important observation on how to construct elements of the Veech group $\Gamma(\omega)$ when the surface ω has a *cylinder decomposition* in a fixed direction with a certain rationality property. Recall from Definition 2.4.2 that a *cylinder* is an isometric image C in ω of

$$\mathcal{C}_{c,h} = \left\{ \left(\frac{c}{2\pi} e^{i\theta}, t \right) : 0 \le \theta < 2\pi, 0 \le t \le h \right\} \subset \mathbb{C} \times \mathbb{R},$$

that the modulus of the cylinder C is defined to be $\mu_C = \frac{h}{c}$, and that cylinders can occur in any fixed direction. Recall from §4.1.1 that we say θ is a *Strebel direction* for ω if the surface decomposes into a union of cylinders $C_1, \ldots, C_k$ and saddle connections on their boundary.

Definition 7.3.1. We say a Strebel direction is *rational* if the moduli $\mu_i = \mu_{C_i}$ have *rational ratios*; that is,

$$\rho_{i,j} := \mu_i / \mu_j \in \mathbb{Q}$$

for $1 \le i, j, \le k$.

Twists. On each cylinder, one can apply a *twist* by the modulus (this corresponds to a *Dehn twist* around the core curve; see [64, §3]). We illustrate how to twist around a *vertical* cylinder by applying the shear

$$u_{\mu^{-1}} = \begin{pmatrix} 1 & 0 \\ \mu^{-1} & 1 \end{pmatrix},$$

in Figure 7.5.

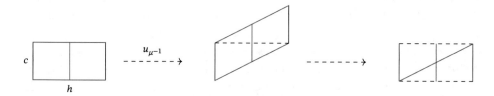

Figure 7.5. Twisting along the core curve (in blue) of a vertical cylinder.

7.4. Optimal dynamics

Common multiple. If a Strebel direction is rational, then there is a minimal common integer multiple $\alpha = n_i \mu_i^{-1}, n_i \in \mathbb{N}$, of the (inverse) moduli of the cylinders. We leave as an exercise:

Exercise 7.7. *With notation as above, show that if the vertical direction on ω is a rational Strebel direction, then*

$$u_\alpha = \begin{pmatrix} 1 & 0 \\ \alpha & 1 \end{pmatrix} \in \Gamma(\omega).$$

Cusps. As discussed in §7.1.3, parabolic elements of a lattice $\Gamma \subset SL(2, \mathbb{R})$ correspond to *cusps* of the quotient $SL(2, \mathbb{R})/\Gamma$, a fact which will be crucial in the next section. In particular, to each cusp of $\Gamma(\omega)$ is associated a parabolic element $\gamma \in \Gamma(\omega)$ and an associated rational Strebel direction for which the cylinder decomposition is stabilized by γ.

7.4. Optimal dynamics

We recall from §1.3 a feature of the linear flow on the torus $(\mathbb{C}/\mathbb{Z}[i], dz)$: in every direction, the flow is either uniquely ergodic or the surface decomposes into periodic orbits (in this case all of the same length). An important property of lattice surfaces is a generalization of this binary known as the *Veech dichotomy*. In both the proof of this theorem and our discussion of counting in §7.5 is the following idea (see, for example, [9, §2.3]): if a property of a translation surface can be interpreted in terms of its $SL(2, \mathbb{R})$-orbit and if ω is a lattice surface, then the property for ω can be studied using the dynamics on the quotient space $SL(2, \mathbb{R})/\Gamma(\omega)$.

The dichotomy. We now state the Veech dichotomy which we will devote the rest of this section to proving.

Theorem 7.4.1 (The Veech dichotomy [164]). *Let ω be a lattice surface. For $\theta \in [0, 2\pi)$, the linear flow in direction θ is either uniquely ergodic or the direction θ is a rational Strebel direction.*

Nonergodicity. In particular, this implies that there are no minimal, non-uniquely ergodic directions on lattice surfaces: if a direction is nonuniquely ergodic, it is a rational Strebel direction.

Divergence. By rotating the surface, we will consider the flow ϕ_t in the vertical direction (note that $\Gamma(r_\theta \omega) = r_\theta \Gamma(\omega) r_{-\theta}$). Suppose under the flow g_t we have that $g_t \omega$ diverges.

Parabolic elements. Since $\Gamma(\omega)$ is not a cocompact lattice, the lattice must contain parabolic elements. (By standard results in hyperbolic geometry (see Dal'Bo [40, §I.2] or S. Katok [100]), $SL(2, \mathbb{R})/\Gamma(\omega)$ consists of a compact part K

and finitely many cusps.) Since $g_t\omega$ diverges, the parabolic element must be of the form
$$u_a = \begin{pmatrix} 1 & 0 \\ a & 1 \end{pmatrix}.$$

Affine maps. Let $\psi \in \mathrm{Aff}^+(\omega)$ be an affine map of ω with $D\psi = u_a$. Then ψ acts by permutation on the set of zeros Σ. By raising ψ to some power j we can assume ψ^j fixes each point of Σ.

Vertical separatrices. At each $p \in \Sigma$ let $\{L_1, L_2, \ldots, L_k\}$ denote the set of outgoing separatrices in the vertical direction. Then ψ^j acts on this set by permutation. Passing to a further power $\psi = \psi^{jm}$ we can assume ψ fixes each point of Σ and fixes the outgoing separatrices at each such point. Now we claim that each L_i is a saddle connection. Since it is fixed as a set and the derivative u_{jma} restricted to the vertical direction is the identity, the map ψ fixes each L_i pointwise. If L_i is not a saddle connection, then it is dense in some open domain and then ψ is the identity in an open set. But this contradicts that its derivative u_{jma} is not the identity. Thus the vertical direction is a Strebel direction.

Rationality. We next prove the direction is a rational Strebel direction. Suppose C_j is a vertical cylinder with circumference c_j, height h_j, and modulus $\mu_j = \frac{h_j}{c_j}$; that is, the closed core curve in the vertical direction has length c_j, and the horizontal width is h_j. Suppose the map ψ consists of n_j twists about the core curve of C_j. Then in the Euclidean coordinates $z = x + iy$ of C_j with $\{iy : 0 \le y \le c_j\}$ coordinates on one boundary component and $\{h_j + iy : 0 \le y \le c_j\}$ on the other we have
$$\psi(z) = x + i\left(y + n_j x \mu_j^{-1} \bmod (c_j)\right)$$
with derivative
$$u_{n_j\mu_j^{-1}} = \begin{pmatrix} 1 & 0 \\ n_j\mu_j^{-1} & 1 \end{pmatrix}.$$
This implies $n_j\mu_j^{-1} = a$ for all j which implies
$$\frac{\mu_k}{\mu_j} = \frac{n_k}{n_j} \in \mathbb{Q}$$
and thus $\frac{\mu_k}{\mu_j} \in \mathbb{Q}$ for all k, j, so the vertical direction is a rational Strebel direction, as desired. We have shown a divergent direction is Strebel. Together with Theorem 5.1.1 this implies that a direction cannot be minimal and not uniquely ergodic. This concludes the proof of Theorem 7.4.1. □

Closed orbits. We have

Lemma 7.4.2. *Let $\omega \in \mathcal{H}$. If ω is a lattice surface, the $SL(2, \mathbb{R})$-orbit of ω is a closed subset of $\mathcal{H}$ and is in bijective correspondence with $SL(2, \mathbb{R})/\Gamma(\omega)$.*

7.4. Optimal dynamics

Proof. Our proof is drawn from the nice survey of Massart [**115**, §7.1]. Note that the orbit map from $o_\omega : SL(2, \mathbb{R}) \to \mathcal{H}$ given by

$$o_\omega(g) \mapsto g\omega$$

descends to a continuous map (which we also denote o_ω) from $SL(2,\mathbb{R})/\Gamma(\omega)$ to $\mathcal{H}$. Then letting $K = SO(2)$ be the maximal compact subgroup of $SL(2, \mathbb{R})$, $o_\omega(K)$ is compact and thus closed in $\mathcal{H}$. Now fix a cusp C of $SL(2,\mathbb{R})/\Gamma(\omega)$. If $o_\omega(C)$ is not closed, there is a sequence $g_n \to \infty$ in C and a compact set $B \subset \mathcal{H}$ with $g_n\omega \in B$. Every cusp is stabilized by a parabolic element $\gamma \in \Gamma(\omega)$, which as we have seen in turn yields a rational Strebel direction of ω. Since $g_n \to \infty$, the lengths of the closed geodesics in this direction must go to 0, and so we leave every compact set in $\mathcal{H}$, a contradiction. $\square$

Smillie's theorem. In fact, the converse of Lemma 7.4.2 is true, a fact originally due to Smillie, whose proof we will sketch in §7.6.

Teichmüller curves. If ω is a lattice surface, the projection of the closed orbit $SL(2, \mathbb{R}) \cdot \omega \cong SL(2,\mathbb{R})/\Gamma(\omega)$ to the moduli space $\mathcal{M}_g$ is an isometric immersion of the hyperbolic surface $\mathbb{H}^2/\Gamma(\omega)$ into $\mathcal{M}_g$ (with respect to the Teichmüller metric on $\mathcal{M}_g$), known as a *Teichmüller curve*. The geometry of these curves in $\mathcal{M}_g$ has been a subject of much interest, which we discuss further in §7.7.

Nonoptimal features. While lattice surfaces share this dichotomy with the torus, there are other important features of dynamics on the torus that are not always reflected for lattice surfaces. In particular, on the torus, long *periodic* trajectories beome equidistributed as their length tends to ∞. Precisely, suppose $f \in C(\mathbb{C}/\mathbb{Z}[i])$ is a mean 0 function; i.e.,

$$\int_{\mathbb{C}/\mathbb{Z}[i]} f\, dm = 0.$$

For a Gaussian integer

$$z = m + ni = re^{i\theta} \in \mathbb{Z}[i],$$

define

$$M_z(f) = \frac{1}{r}\int_0^r f(te^{i\theta})dt$$

to be the average of f along the periodic geodesic associated to z. We leave as an exercise:

Exercise 7.8. *With notation as above, show*

$$\lim_{|z|\to\infty} M_z(f) = 0,$$

where the limit is taken for $z \in \mathbb{Z}[i]$.

Pentagons. Motivated by problems of billiards, Davis-Lelièvre [43] showed that for a lattice translation surface associated to billiards in a regular pentagon, long periodic trajectories do not neccessarily equidistribute, a fact explained in more detail by McMullen [127].

7.5. Homogeneous dynamics

We now turn from dynamics to *counting* and show how the strategy described for counting saddle connections in Chapter 6 becomes simpler in the context of lattice surfaces, a fact exploited by Veech [165, §16]. In particular, we will be able to replace the role played by Nevo's ergodic theorem, Theorem 6.5.1, with a classical result on equidistribution of large circles on $SL(2, \mathbb{R})/\Gamma(\omega)$.

7.5.1. Holonomy vectors for lattice surfaces. We first record a result of Veech [164] that the set of holonomy vectors of saddle connections Λ_ω for a lattice surface can be described in terms of orbits of the Veech group $\Gamma(\omega)$ acting $\mathbb{R}$-linearly on $\mathbb{C}^*$; namely, Λ_ω is a *finite union* of $\Gamma(\omega)$-orbits on $\mathbb{C}^*$.

Theorem 7.5.1 ([164])**.** *Let ω and Γ be as above. There are $z_1, \ldots, z_N \in \mathbb{C}^*$ such that*

$$\Lambda_\omega = \bigsqcup_{i=1}^{N} \Gamma z_i.$$

Linear action of $SL(2, \mathbb{Z})$. Note that $SL(2, \mathbb{Z})$ has one cusp. A representative maximal parabolic subgroup is

$$\Gamma_\infty = \left\{ h_n = \begin{pmatrix} 1 & n \\ 0 & 1 \end{pmatrix} : n \in \mathbb{Z} \right\}$$

stabilizing ∞ under the fractional linear action

$$\begin{pmatrix} a & b \\ c & d \end{pmatrix} \cdot x = \frac{ax + b}{cx + d}$$

on $\partial \mathbb{H}^2 = \mathbb{R} \cup \infty$. Any other maximal parabolic subgroup of $SL(2, \mathbb{Z})$ is conjugate to Γ and stabilizes some $q \in \mathbb{Q} \cup \infty$. If we consider the $\mathbb{R}$-linear action of $SL(2, \mathbb{Z})$ on $\mathbb{C}^*$, we have that orbits are either *discrete* or *dense*. For example, note that $\mathbb{Z}_{\text{prim}}[i] = SL(2, \mathbb{Z}) \cdot 1$. More generally, $SL(2, \mathbb{Z}) \cdot z$ is discrete if and only if $z = x + iy$ with $x/y \in \mathbb{Q}$, in which case there is a maximal parabolic subgroup of Γ which fixes z. We leave the proof as

Exercise 7.9. *Let $z = x + iy \in \mathbb{C}$. Show that if $x/y \in \mathbb{Q}$, then $SL(2, \mathbb{Z}) \cdot z$ is discrete in $\mathbb{C}^*$, and if $x/y \notin \mathbb{Q}$, then $SL(2, \mathbb{Z}) \cdot z$ is dense in $\mathbb{C}^*$.*

7.5. Homogeneous dynamics

Linear actions of Fuchsian groups. More generally, if Γ is a lattice in $SL(2, \mathbb{R})$, we have the following generalization (see, for example, Ledrappier [113]) of Exercise 7.9. Let P_Γ denote the set of fixed points of maximal parabolic subgroups of Γ. If $z = x + iy \in \mathbb{C}$ has $z = x/y \in P_\Gamma$, then z is fixed by a maximal parabolic subgroup of Γ under the linear action of Γ on $\mathbb{C}$.

Black Box 7.5.2. *Let $z = x + iy \in \mathbb{C}$. If $x/y \in P_\Gamma$, then $\Gamma \cdot z$ is discrete in $\mathbb{C}^*$, and if $x/y \notin P_\Gamma$, then $\Gamma \cdot z$ is dense in $\mathbb{C}^*$.*

Horocycle flow. This fact is closely related to the result of Dani-Smillie [41], mentioned in §5.3.2, that for Γ a nonuniform lattice, horocycle flow orbits on $SL(2, \mathbb{R})/\Gamma$ are either closed (corresponding to cusps and discrete Γ-orbits on $\mathbb{C}^*$) or equidistributed and thus dense (corresponding to dense Γ-orbits on $\mathbb{C}^*$). This is because setting $N = \{h_s : s \in \mathbb{R}\}$ to be the 1-parameter subgroup of $SL(2, \mathbb{R})$ whose action gives the horocycle flow, we have

$$SL(2, \mathbb{R})/N \cong \mathbb{C}^*,$$

since the $\mathbb{R}$-linear $SL(2, \mathbb{R})$-action on $\mathbb{C}^*$ is transitive and the stabilizer of 1 is N. So both Black Box 7.5.2 and the Dani-Smillie result can be viewed as statements about double cosets $Ng\Gamma$. That is, we can go between the linear Γ-action on $SL(2, \mathbb{R})/N = \mathbb{C}^*$ and the N-action (horocycle flow) on $SL(2, \mathbb{R})/\Gamma$.

Cusps and Strebel directions. We now prove Theorem 7.5.1. Suppose ω is a lattice surface. Note that by construction, Λ_ω is $\Gamma(\omega)$-invariant. Now let $z \in \Lambda_\omega$. Then there is a maximal parabolic subgroup (that is, a representative of a cusp) of $\Gamma(\omega)$ fixing z. There are only finitely many cusps $\mathcal{C}_1, \ldots, \mathcal{C}_n$ of $\Gamma(\omega)$, that is, only finitely many conjugacy classes of maximal parabolic subgroups. For each cusp $\mathcal{C}_i$, there is a $\Gamma(\omega)$-indexed collection of slopes fixed by $\Gamma(\omega)$-conjugates. Each of these corresponds to a $\Gamma(\omega)$-orbit on $\mathbb{R} \cup \infty$, which corresponds to a set of rational Strebel directions for ω. In each direction, there are a finite number of saddle connections $\gamma_{i,1}, \ldots, \gamma_{i,k_i}$. Let

$$z_{i,j} = \int_{\gamma_{i,j}} \omega.$$

Then we must have $z = g \cdot z_{i,j}$ for some $g \in \Gamma(\omega)$, and so we have expressed Λ_ω as a finite union of $\Gamma(\omega)$-orbits, proving Theorem 7.5.1. $\square$

Explicit computations. The description of Λ_ω as a finite union of $\Gamma(\omega)$-orbits can allow one to explicitly compute these sets of holonomy vectors using the generators of $\Gamma(\omega)$. Such a strategy, inspired by the Farey tree structure of $SL(2, \mathbb{Z})$ and related structures on $\Delta(2, 5, \infty)$, was used for explicit computations on the golden L in [9] and subsequently for translation surfaces with Veech group $\Delta(2, q, \infty)$ by Taha [158].

7.5.2. Counting in orbits.
Theorem 7.5.1 reduces the asymptotic counting problem of
$$N(\omega, R) = \#\Lambda_\omega \cap B(0, R)$$
for lattice surfaces $\Gamma(\omega)$ to a problem about orbits of Fuchsian groups. Namely, if Γ is a nonuniform lattice in $SL(2, \mathbb{R})$ and $z \in \mathbb{C}^*$ has a discrete Γ-orbit, can we understand the asymptotics of
$$N_\Gamma(z, R) = \#\left((\Gamma \cdot z) \cap B(0, R)\right)?$$
Veech [164] used Eisenstein series ideas to address this problem and, subsequently, in [165], used dynamics. The counting strategy described in Chapter 6 can once again be implemented, with $\mathcal{H}$ replaced by $SL(2, \mathbb{R})/\Gamma$ and the equivariant assignment
$$g\Gamma \mapsto g\Gamma \cdot z \subset \mathbb{C}^*.$$
The problem of counting reduces to understanding the limiting behavior of the operators A_τ acting on functions on $SL(2, \mathbb{R})/\Gamma$. The mixing of the geodesic flow on $SL(2, \mathbb{R})/\Gamma$ can be used (see, for example, Eskin-McMullen [60] or Bekka-Mayer [25] for nice expositions) to conclude that the operators A_τ of a function converge, as $\tau \to \infty$, to averaging the function over $SL(2, \mathbb{R})/\Gamma$ with respect to Haar measure. This yields the following result of Veech:

Theorem 7.5.3. *With notation as above, there is a $c = c(\Gamma, z)$ such that*
$$\lim_{R \to \infty} \frac{N_\Gamma(z, R)}{R^2} = c(\Gamma, v).$$

Branched covers of lattice surfaces. We note that while translation covers of lattice surfaces (that is, covers which are branched only over zeros) yield lattice surfaces, covers branched over other points need not be lattice surfaces. Motivated by problems on billiards and using ideas from homogeneous dynamics, Eskin-Marklof-Morris [56] were able to obtain limit theorems for the operators A_τ on certain subvarieties of $\mathcal{H}$ to obtain counting results for particular families of surfaces arising as branched covers of lattice surfaces.

7.6. Characterizations of lattice surfaces

There are many interesting equivalent ways to characterize lattice surfaces. We collect several examples of these results, sketching proofs when the arguments are within the scope of this book.

7.6.1. Closed orbits.
We showed above in §7.3 that if $\omega \in \mathcal{H}$ is a lattice surface, $SL(2, \mathbb{R})/\Gamma(\omega)$ is a closed subset of $\mathcal{H}$. In fact, the converse is also true. This was originally due to an unpublished argument of Smillie. We sketch

7.6. Characterizations of lattice surfaces

a proof following the exposition in Gekhtman-Wright [78], which uses the Minsky-Weiss nondivergence result Theorem 5.3.4.

Theorem 7.6.1. *Let $\omega \in \mathcal{H}$. Suppose $SL(2,\mathbb{R}) \cdot \omega$ is a closed subset of $\mathcal{H}$. Then ω is a lattice surface and the orbit can be identified with $SL(2,\mathbb{R})/\Gamma(\omega)$.*

Sketch of a proof. First, since the orbit is closed, it can be identified as an immersed copy (that is, the image under an immersion) of $SL(2,\mathbb{R})/\Gamma(\omega)$, the unit tangent bundle to a hyperbolic orbifold. Now, using the Minsky-Weiss nondivergence result Theorem 5.3.4, the horocycle flow is nondivergent. Standard arguments from homogeneous dynamics imply that if the horocycle flow on the unit tangent bundle of a hyperbolic orbifold $SL(2,\mathbb{R})/\Gamma$ is nondivergent in the sense of Theorem 5.3.4, the orbifold $SL(2,\mathbb{R})/\Gamma$ must have finite hyperbolic volume; that is, $\Gamma(\omega)$ must be a lattice. □

7.6.2. Rationally related moduli.
Another necessary condition for ω to be a lattice surface that we saw in §7.4 is that every nonuniquely ergodic direction was in fact a rational Strebel direction. Vorobets [168] showed that a slight strengthening of this property is also a sufficient condition, namely:

Theorem 7.6.2 ([168, Proposition 6.1]). *Let $\omega \in \mathcal{H}$. Then ω is a lattice surface if and only if every nonuniquely ergodic direction is a rational Strebel direction and the ratios of lengths of saddle connections in a fixed direction is* uniformly bounded *(as the direction varies).*

Weakened Veech dichotomy. On the other hand Smillie-Weiss [156] showed that the surfaces ω constructed by Hubert-Schmidt [93] with $\Gamma(\omega)$ infinitely generated satisfy a weakened Veech dichotomy, namely, that every nonuniquely ergodic direction is a Strebel direction. Recall that the fact that $\Gamma(\omega)$ is infinitely generated means ω is not a lattice surface.

7.6.3. No small triangles.
Vorobets [168] gave other geometric necessary conditions for a surface to be a lattice surface, in terms of *triangles* on ω. This condition, known as the *no small triangles* condition, was shown to be *sufficient* by Smillie-Weiss [155].

Definition 7.6.3. Let $\omega \in \mathcal{H}$. A *triangle* in ω is the image of a triangle in $\mathbb{C}$ under a map which is isometric on the interior, with vertices mapped into the set Σ of zeros of ω. The images of the vertices need not be distinct.

Areas. The area of a triangle is thus well-defined, and we denote the set of areas of triangles in ω by $\mathcal{T}(\omega)$. We have the following, by results of Vorobets [168] and Smillie-Weiss [155]:

Theorem 7.6.4. *Let $\omega \in \mathcal{H}$. The following three conditions are equivalent:*

Finitely many triangles: *The set $\mathcal{T}(\omega)$ is finite.*

No small triangles: *The set $\mathcal{T}(\omega)$ is bounded away from 0; that is,*
$$\inf \mathcal{T}(\omega) > 0.$$

Lattice surface: *ω is a lattice surface.*

Equivalences. Vorobets [168] showed that the lattice property implies the first two properties and the lattice property is equivalent to the finiteness of $\mathcal{T}(\omega)$. He showed that the no small triangles property implied that ω satisfies the Veech dichotomy (that is, the conclusion of Theorem 7.4.1) and asked whether the no small triangles property was in fact equivalent to the lattice property. This final implication, of the no small triangles property implying the lattice property, was shown by Smillie-Weiss [155], who gave several other equivalent characterizations of the lattice property.

No small virtual triangles. We mention one of these characterizations, which is purely in terms of the set of holonomy vectors Λ_ω. Given $z = x + iy, w = u + iv \in \mathbb{C}^*$, let
$$|z \wedge w| = |xv - yu|$$
be the area of the parallelogram spanned by z, w. We refer to a pair of (non-parallel) saddle connections γ_1, γ_2 as a *virtual triangle*, and we define the virtual area to be the area $|z_{\gamma_1} \wedge z_{\gamma_2}|$ of the parallelogram spanned by their holonomy vectors. Let
$$\mathcal{VT}(\omega) := \{|z \wedge w| : z, w \in \Lambda_\omega, z/w \notin \mathbb{R}\} \subset \mathbb{R}^+$$
be the set of nonzero virtual areas of holonomy vectors of saddle connections on ω. Smillie-Weiss [155] showed the following:

Theorem 7.6.5. *Let $\omega \in \mathcal{H}$. Then ω is a lattice surface if and only if it has the* no small virtual triangles *property; that is,*
$$\inf \mathcal{VT}(\omega) > 0.$$

7.7. Classifying lattice surfaces

The problem of *classifying* lattice surfaces has generated a huge amount of interest from a wide variety of perspectives. We recommend the comprehensive survey of McMullen [128] for a thorough description of what is known and for connections to Teichmüller theory, algebraic geometry, and number theory. Drawing on this survey, we mention a few important definitions and theorems (without proof).

7.7. Classifying lattice surfaces

7.7.1. Primitivity. We have seen in §7.2.5 that taking branched covers of lattice surfaces branched over zeros yields new lattice surfaces, so to consider a well-defined classification problem, we need a notion of *primitivity*.

Definition 7.7.1. We say $\omega \in \mathcal{H}$ (of genus $g > 1$) is *primitive* if it is not the pullback of a form of lower genus.

Primitive forms. We have (see, for example, McMullen [**124**, Theorem 2.1]):

Black Box 7.7.2. *Let $\omega \in \mathcal{H}$ be of genus $g > 1$. Then ω is the pullback of a unique primitive η (if ω is itself primitive, $\omega = \eta$).*

Affine groups and Veech groups. For a primitive lattice surface ω, we have
$$\mathrm{Aff}^+(\omega) = \Gamma(\omega),$$
since the subgroup of the automorphism group $\mathrm{Aut}(X)$ which preserves ω is trivial by the primitivity property.

7.7.2. Trace field. There are several important invariants associated to lattice surfaces. We mention a number-theoretic invariant, starting with the *trace field*.

Definition 7.7.3. Let ω be a lattice surface. Define the *trace field* $K(\omega)$ to be the extension of $\mathbb{Q}$ obtained by adjoining traces of elements of $\Gamma(\omega)$:
$$K(\omega) := \mathbb{Q}(\{\mathrm{Tr}(g) : g \in \Gamma(\omega)\}).$$

Properties. The trace field has many nice properties. We have (see, for example, Hubert-Lanneau [**90**] and [**128**, §2])

Theorem 7.7.4 ([**90**, Theorem 1.1]). *Let ω be a lattice surface of genus g. $K(\omega)$ is totally real and has degree at most g over $\mathbb{Q}$, and*
$$K(\omega) = \mathbb{Q}(\mathrm{Tr}(g))$$
for any hyperbolic element $g \in \Gamma(\omega)$.

7.7.3. The Jacobian. An important invariant of a Riemann surface X is the *Jacobian* $J(X)$, a complex g-dimensional torus (see, for example, Jost [**95**, §5.3]). Recall that $\Omega(X)$ denotes the vector space of holomorphic 1-forms on X, which has complex dimension g. Fix a basis $\omega_1, \ldots, \omega_g$ for $\Omega(X)$ and basis $a_1, \ldots, a_g$, $b_1, \ldots, b_g$ for homology $H_1(X)$. Form the *period matrix* $P(X)$ whose ith row is given by integrating ω_i over the curves $a_1, \ldots, a_g, b_1, \ldots, b_g$. The Jacobian
$$J(X) = \Omega^*(X)/P(X)\mathbb{Z}^{2g}$$
is a complex torus, which does not depend on the choices of basis and is in fact a *principally polarized Abelian variety*. A geometric view of polarization is the choice of a *symplectic form* on $H_1(X, \mathbb{R})$ given by the intersection pairing.

Traces, Jacobians, and real multiplication. We present a result of Möller [**131**] connecting lattice surfaces to properties of Jacobians, without proof. We refer to [**128**, §1] for our version of this statement and to [**128**, §4] for details on real multiplication.

Theorem 7.7.5. *Let ω be a primitive lattice surface with underlying Riemann surface X, with Veech group $\Gamma = \Gamma(\omega)$ and trace field $K = K(\omega)$. Then*
$$\Gamma \subset SL(2, K),$$
ω can be assembled from triangles with vertices in $\{k_1 + k_2\tau : k_i \in K\}$ for some $\tau \in \mathbb{H}^2$, and there is a factor of the Jacobian $J(X)$ which admits a natural injection from K into its set of self-adjoint endomorphisms.

7.7.4. Genus 2 surfaces. To give an idea of the power of the ideas in Theorem 7.7.5, we describe a classification result of McMullen [**128**, Corollary 4.11], who showed, using real muliplication ideas, that all Teichmüller curves that come from lattice surfaces in $\mathcal{H}(2)$ arise from particular values of the family $L(a, b)$ shown in Figure 7.2:

Theorem 7.7.6 ([**128**, Corollary 4.11]). *The translation surface $\omega_{L(a,b)}$ is a lattice surface if and only if $a(b - 1), b(a - 1) \in \mathbb{Q}$. For example, if $a = b$, we must have $a = (1 + \sqrt{k})/2$ for some rational $k > 1$. All lattice surfaces in $\mathcal{H}(2)$ arise in this manner.*

Calta's results. Calta [**31**], using different methods, independently proved Theorem 7.7.6, while classifying completely periodic surfaces (those for which all nonuniquely ergodic directions are Strebel) in $\mathcal{H}(2)$ (showing that they are all lattice surfaces) and in $\mathcal{H}(1, 1)$ (finding explicit nonlattice examples).

7.7.5. Other classification results. We mentioned the constructions of Veech [**164**], Ward [**169**], Bouw-Möller [**28**], and Hooper [**85**] in §7.2.3. In addition to these constructions which give infinite families of primitive Teichmüller curves with growing genus, there are explicit constructions of infinite families of primitive lattice surfaces in genus 3 and 4. There are many contributors to this story, and we refer the interested reader to McMullen's survey [**128**] for details. A currently open question is whether there are infinitely many primitive lattice surfaces in genus 5. We note that the Bouw-Möller construction (which includes the Veech and Ward constructions as special cases) is the only known construction of primitive lattice surfaces in genus $g \geq 5$.

7.8. Square-tiled surfaces

We now turn to a brief discussion of the subject of *square-tiled surfaces*, that is, translation covers of the torus. We will follow the notes of Matheus [**122**] and the survey of Schmithüsen [**149**] closely, and we urge the interested reader to

7.8. Square-tiled surfaces

pursue those surveys and the references therein to develop a better idea of the properties of this important family of surfaces. We note that the terminology *origami* is often used for these surfaces. We recall that a square-tiled surface (or origami) is a translation surface ω which is a pullback of the 1-form dz on $\mathbb{C}/\mathbb{Z}[i]$, branched over 0, which we can view as a surface tiled by squares. As we discussed in §7.2.1, these are all *lattice surfaces*, with Veech groups $\Gamma(\omega)$ commensurable to $SL(2,\mathbb{Z})$.

7.8.1. Permutations and monodromy. There is a purely *combinatorial* definition of square-tiled surfaces: fix two permutations σ_h, σ_v in the symmetric group S_n (h for horizontal, v for vertical) such that $\langle \sigma_h, \sigma_v \rangle$ acts transitively on $\{1, \ldots, n\}$. Form a translation surface $\omega_{\sigma_h, \sigma_v}$ by taking n squares indexed by $\{1, \ldots, n\}$ and gluing the top (horizontal) side of square i to the bottom of square $\sigma_h(i)$ and gluing the right (vertical) side of square i to the left side of square $\sigma_v(i)$. We leave as an (easy) exercise:

Exercise 7.10. *Give permutations $\sigma_h, \sigma_v \in S_3$ that yield the 3-square L as in Figure 7.1.*

Transitivity. The transitivity of $\langle \sigma_h, \sigma_v \rangle$ guarantees that the resulting surface is connected, a proof which we leave as an exercise:

Exercise 7.11. *With notation as above, show that $\omega_{\sigma_h, \sigma_v}$ is connected.*

Conjugation. If $\tau \in S_n$, the surfaces $\omega_{\sigma_h, \sigma_v}$ and $\omega_{\tau \sigma_h \tau^{-1}, \tau \sigma_v \tau^{-1}}$ are clearly the same surface (only the numbering of the squares changes). To determine a branched cover of the torus over 0 with degree n, we have a representation of the free group F_2, which is the fundamental group of the punctured torus, into the potential deck group S_n, and we can think of σ_h and σ_v as images of the generators of F_2. When building origamis, we are considering these representations up to conjugation.

Regular origamis. The permutation definition allows an elegant group-theoretic construction of origamis, known as *regular origamis*.

Definition 7.8.1. Let $G = \langle v, h \rangle$ be a finite group, with cardinality $|G| = n$ generated by two elements v and h. The *regular origami* $\omega_{(G,v,h)}$ is given by n squares indexed by elements of G, with the right side of the square labeled by g glued to the left side of gv and with the top of the square labeled by g glued to the bottom of gh.

Connectedness. Note that regular origamis are connected by construction, since $G = \langle v, h \rangle$.

Eierlegende Wollmichsau. If we set the G to be the quaternion group $G = \{\pm 1, \pm i, \pm j, \pm k\}$, with $i^2 = j^2 = k^2 = -1$, $ij = k$, $jk = i$, $ki = j$, and if we set $v = i, h = j$ (check that these generate G!), the associated regular origami is the so-called *Eierlegende Wollmilchsau*, which we denote ω_{EW}. The name comes from a legendary fictional German animal, the "egg-laying, wool and milk giving pig", referring to the fact that this surface has been a counterexample to many conjectures about translation surfaces.

Exercise 7.12. *Draw a picture of ω_{EW}, and show $\omega_{EW} \in \Omega(1,1,1,1)$.*

$SL(2,\mathbb{Z})$-**action.** The group $SL(2,\mathbb{Z})$ acts on each square of an origami, and the resulting surface can be reassembled into a new origami (since $SL(2,\mathbb{Z})$ acts on the torus). Recall that $SL(2,\mathbb{Z})$ is generated by

$$h_1 = \begin{pmatrix} 1 & 1 \\ 0 & 1 \end{pmatrix}, \quad u_1 = \begin{pmatrix} 1 & 0 \\ 1 & 1 \end{pmatrix}.$$

We leave as an exercise:

Exercise 7.13. *Show that*

$$h_1 \omega_{\sigma_h, \sigma_v} = \omega_{\sigma_h, \sigma_v \sigma_h^{-1}}, \quad u_1 \omega_{\sigma_h, \sigma_v} = \omega_{\sigma_h \sigma_v^{-1}, \sigma_v}.$$

Finiteness of orbits. Following Matheus [122], we observe that the $SL(2,\mathbb{Z})$-orbit of any origami is finite: if the origami has n squares, the size of the orbit is bounded by $|S_n|^2 = (n!)^2$. This implies part of the Gutkin-Judge result Black Box 7.2.3 that if ω is an origami, the Veech group $\Gamma(\omega)$ is a finite-index subgroup of $SL(2,\mathbb{Z})$, since $\Gamma(\omega)$ is the stabilizer of ω.

7.8.2. Integer points. We now turn to the idea that square-tiled surfaces are *integer points* in our period coordinates on strata. Note that if we do not normalize the area (that is, we think of a surface with n squares as having area n), the periods of a square-tiled surface will be in $\mathbb{Z}[i]$ (and if a surface has periods in $\mathbb{Z}[i]$, it is square-tiled; see, for example, [81, Theorem 5.5]). Note that since changes in period coordinates are by integral symplectic matrices, this notion does not change under coordinate changes. Note that by projecting these surfaces to $\mathcal{H}$, one obtains a *dense* collection of square-tiled surfaces, showing that lattice surfaces are dense in any stratum.

Counting and volumes. Following Eskin's survey [53, §2], we sketch an argument about how computing volumes $\mu_{\mathcal{H}}(\mathcal{H})$ can be related to a problem of counting square-tiled surfaces. Recall the *cone construction* (3.4.1), (3.4.2) of the measure $\mu_{\mathcal{H}}$ from the measure ν on Ω from §3.4. We have

$$\mu_{\mathcal{H}}(\mathcal{H}) = \nu(C(\mathcal{H})).$$

For $R > 0$, let

$$C_R(\mathcal{H}) = \{r\omega : \omega \in \mathcal{H}, 0 < r \leq R\}.$$

Then

$$\nu(C_R(\mathcal{H})) = R^k \nu(C(\mathcal{H})) = R^k \mu_{\mathcal{H}}(\mathcal{H}),$$

7.8. Square-tiled surfaces

where $k = 2N = 2(2g + s - 1)$ is the real dimension of Ω. Since ν is built from Lebesgue measure, we have that the volume $\nu(C_R(\mathcal{H}))$ is asymptotic to the number of integer points in $C_R(\mathcal{H})$, which correspond exactly to square-tiled surfaces in Ω tiled by at most R^2 unit-area squares (since $\text{Area}(r\omega) = r^2 \text{Area}(\omega)$, the surfaces in $C_R(\mathcal{H})$ have area at most R^2).

Quasimodularity. We denote this count by $N_\Omega(R)$. This counting was undertaken by Eskin-Okounkov [59], who showed a deep connection to number theory, in particular proving a conjecture of Kontsevich that the generating function

$$F_\Omega(q) = \sum_{d \geq 1} N_\Omega(d) q^d$$

is a *quasimodular form*, a polynomial in the classical Eisenstein series $G_w(q)$, $w = 2, 4, 6$. As a consequence, they showed

$$\mu_\mathcal{H}(\mathcal{H}) \pi^{-2g} \in \mathbb{Q},$$

where g is the genus of surfaces in $\mathcal{H}$. This circle of ideas has been developed further by, among others, Chen-Möller-Zagier [35], who develop further connections to quasimodularity, and by Delecroix-Goujard-Zograf-Zorich [44] who show *equidistribution* properties for the set of square-tiled surfaces.

Conclusion

As we discussed in the preface, this is a very selective account of the theory of translation surfaces and its connections to many different areas of mathematics. By the nature of the book-writing and production process and by the current speed of development in the field, there will be many new and interesting results on translation surfaces that will have appeared before this book does. We hope that reading this book (and doing the exercises!) prepares the interested reader to appreciate, keep abreast of, and contribute to the development of this exciting field.

Bibliography

[1] Ralph Abraham and Jerrold E. Marsden, *Foundations of mechanics*, 2nd ed., revised and enlarged; with the assistance of Tudor Raţiu and Richard Cushman, Benjamin/Cummings Publishing Co., Inc., Advanced Book Program, Reading, MA, 1978. MR515141

[2] Lars V. Ahlfors, *Lectures on quasiconformal mappings*, 2nd ed., with supplemental chapters by C. J. Earle, I. Kra, M. Shishikura, and J. H. Hubbard, University Lecture Series, vol. 38, American Mathematical Society, Providence, RI, 2006, DOI 10.1090/ulect/038. MR2241787

[3] Pierre Arnoux, *Le codage du flot géodésique sur la surface modulaire* (French, with English and French summaries), Enseign. Math. (2) **40** (1994), no. 1-2, 29–48. MR1279059

[4] Pierre Arnoux and Jean-Christophe Yoccoz, *Construction de difféomorphismes pseudo-Anosov* (French, with English summary), C. R. Acad. Sci. Paris Sér. I Math. **292** (1981), no. 1, 75–78. MR610152

[5] Michael Artin, *Algebra*, Prentice Hall, Inc., Englewood Cliffs, NJ, 1991. MR1129886

[6] Jayadev S. Athreya, David Aulicino, and W. Patrick Hooper, *Platonic solids and high genus covers of lattice surfaces*, with an appendix by Anja Randecker, Experimental Mathematics **31** (2022), no. 3, 847–877, DOI 10.1080/10586458.2020.1712564. MR4477409

[7] J. S. Athreya and J. Chaika, *The distribution of gaps for saddle connection directions*, Geom. Funct. Anal. **22** (2012), no. 6, 1491–1516, DOI 10.1007/s00039-012-0164-9. MR3000496

[8] Jayadev S. Athreya and Jon Chaika, *The Hausdorff dimension of non-uniquely ergodic directions in $H(2)$ is almost everywhere $\frac{1}{2}$*, Geom. Topol. **19** (2015), no. 6, 3537–3563, DOI 10.2140/gt.2015.19.3537. MR3447109

[9] Jayadev S. Athreya, Jon Chaika, and Samuel Lelièvre, *The gap distribution of slopes on the golden L*, Recent trends in ergodic theory and dynamical systems, Contemp. Math., vol. 631, Amer. Math. Soc., Providence, RI, 2015, pp. 47–62, DOI 10.1090/conm/631/12595. MR3330337

[10] Jayadev S. Athreya, Yitwah Cheung, and Howard Masur, *Siegel-Veech transforms are in L^2*, with an appendix by Athreya and Rene Rühr, J. Mod. Dyn. **14** (2019), 1–19, DOI 10.3934/jmd.2019001. MR3959354

[11] Jayadev S. Athreya, Alex Eskin, and Anton Zorich, *Right-angled billiards and volumes of moduli spaces of quadratic differentials on $\mathbb{C}P^1$* (English, with English and French summaries), with an appendix by Jon Chaika, Ann. Sci. Éc. Norm. Supér. (4) **49** (2016), no. 6, 1311–1386, DOI 10.24033/asens.2310. MR3592359

[12] J. S. Athreya, S. Fairchild, and H. Masur, *Counting pairs of saddle connections*, Adv. Math. **431** (2023), Paper No. 109233, 55, DOI 10.1016/j.aim.2023.109233. MR4624870

[13] J. S. Athreya and G. Forni, *Deviation of ergodic averages for rational polygonal billiards*, Duke Math. J. **144** (2008), no. 2, 285–319, DOI 10.1215/00127094-2008-037. MR2437681

[14] David Aulicino, *Teichmüller discs with completely degenerate Kontsevich-Zorich spectrum*, Comment. Math. Helv. **90** (2015), no. 3, 573–643, DOI 10.4171/CMH/365. MR3420464

[15] David Aulicino, *Affine manifolds and zero Lyapunov exponents in genus 3*, Geom. Funct. Anal. **25** (2015), no. 5, 1333–1370, DOI 10.1007/s00039-015-0339-2. MR3426056

[16] David Aulicino, *Affine invariant submanifolds with completely degenerate Kontsevich-Zorich spectrum*, Ergodic Theory Dynam. Systems **38** (2018), no. 1, 10–33, DOI 10.1017/etds.2016.26. MR3742536

[17] Artur Avila and Vincent Delecroix, *Weak mixing directions in non-arithmetic Veech surfaces*, J. Amer. Math. Soc. **29** (2016), no. 4, 1167–1208, DOI 10.1090/jams/856. MR3522612

[18] Artur Avila, Alex Eskin, and Martin Möller, *Symplectic and isometric SL(2, $\mathbb{R}$)-invariant subbundles of the Hodge bundle*, J. Reine Angew. Math. **732** (2017), 1–20, DOI 10.1515/crelle-2014-0142. MR3717086

[19] Artur Avila and Giovanni Forni, *Weak mixing for interval exchange transformations and translation flows*, Ann. of Math. (2) **165** (2007), no. 2, 637–664, DOI 10.4007/annals.2007.165.637. MR2299743

[20] Artur Avila, Giovanni Forni, and Pedram Safaee, *Quantitative weak mixing for interval exchange transformations*, Geom. Funct. Anal. **33** (2023), no. 1, 1–56, DOI 10.1007/s00039-023-00625-y. MR4561147

[21] Artur Avila and Sébastien Gouëzel, *Small eigenvalues of the Laplacian for algebraic measures in moduli space, and mixing properties of the Teichmüller flow*, Ann. of Math. (2) **178** (2013), no. 2, 385–442, DOI 10.4007/annals.2013.178.2.1. MR3071503

[22] Artur Avila, Sébastien Gouëzel, and Jean-Christophe Yoccoz, *Exponential mixing for the Teichmüller flow*, Publ. Math. Inst. Hautes Études Sci. **104** (2006), 143–211, DOI 10.1007/s10240-006-0001-5. MR2264836

[23] Artur Avila and Marcelo Viana, *Simplicity of Lyapunov spectra: proof of the Zorich-Kontsevich conjecture*, Acta Math. **198** (2007), no. 1, 1–56, DOI 10.1007/s11511-007-0012-1. MR2316268

[24] Matt Bainbridge, Dawei Chen, Quentin Gendron, Samuel Grushevsky, and Martin Möller, *Compactification of strata of Abelian differentials*, Duke Math. J. **167** (2018), no. 12, 2347–2416, DOI 10.1215/00127094-2018-0012. MR3848392

[25] M. Bachir Bekka and Matthias Mayer, *Ergodic theory and topological dynamics of group actions on homogeneous spaces*, London Mathematical Society Lecture Note Series, vol. 269, Cambridge University Press, Cambridge, 2000, DOI 10.1017/CBO9780511758898. MR1781937

[26] Corentin Boissy, *Connected components of the strata of the moduli space of meromorphic differentials*, Comment. Math. Helv. **90** (2015), no. 2, 255–286, DOI 10.4171/CMH/353. MR3351745

[27] Etienne Bonnafoux, *Pairs of saddle connections of typical flat surfaces on fixed affine orbifolds*, Preprint, arXiv:2209.11862, 2022.

[28] Irene I. Bouw and Martin Möller, *Teichmüller curves, triangle groups, and Lyapunov exponents*, Ann. of Math. (2) **172** (2010), no. 1, 139–185, DOI 10.4007/annals.2010.172.139. MR2680418

[29] Joshua P. Bowman, *The complete family of Arnoux-Yoccoz surfaces*, Geom. Dedicata **164** (2013), 113–130, DOI 10.1007/s10711-012-9762-9. MR3054619

[30] Andrei Bud and Chen Dawei, *Moduli of differentials and Teichmüller dynamics* (2022), https://drive.google.com/file/d/16Ry9_oTZQj9WfiMwaRgi9bEedOv8WZRH/view.

[31] Kariane Calta, *Veech surfaces and complete periodicity in genus two*, J. Amer. Math. Soc. **17** (2004), no. 4, 871–908, DOI 10.1090/S0894-0347-04-00461-8. MR2083470

[32] Jon Chaika and Howard Masur, *The set of non-uniquely ergodic d-IETs has Hausdorff codimension 1/2*, Invent. Math. **222** (2020), no. 3, 749–832, DOI 10.1007/s00222-020-00978-3. MR4169051

[33] Jon Chaika, John Smillie, and Barak Weiss, *Tremors and horocycle dynamics on the moduli space of translation surfaces*, arXiv:2004.04027.

[34] Dawei Chen and Quentin Gendron, *Towards a classification of connected components of the strata of k-differentials*, Doc. Math. **27** (2022), 1031–1100. MR4452231

[35] Dawei Chen, Martin Möller, and Don Zagier, *Quasimodularity and large genus limits of Siegel-Veech constants*, J. Amer. Math. Soc. **31** (2018), no. 4, 1059–1163, DOI 10.1090/jams/900. MR3836563

[36] Yitwah Cheung, *Hausdorff dimension of the set of nonergodic directions*, with an appendix by M. Boshernitzan, Ann. of Math. (2) **158** (2003), no. 2, 661–678, DOI 10.4007/annals.2003.158.661. MR2018932

[37] Yitwah Cheung, Pascal Hubert, and Howard Masur, *Dichotomy for the Hausdorff dimension of the set of nonergodic directions*, Invent. Math. **183** (2011), no. 2, 337–383, DOI 10.1007/s00222-010-0279-2. MR2772084

[38] John B. Conway, *Functions of one complex variable*, 2nd ed., Graduate Texts in Mathematics, vol. 11, Springer-Verlag, New York-Berlin, 1978. MR503901

[39] Yves Coudène, *Ergodic theory and dynamical systems*, translated from the 2013 French original [MR3184308] by Reinie Erné, Universitext, Springer-Verlag London, Ltd., London; EDP Sciences, [Les Ulis], 2016, DOI 10.1007/978-1-4471-7287-1. MR3586310

[40] Françoise Dal'Bo, *Geodesic and horocyclic trajectories*, translated from the 2007 French original, Universitext, Springer-Verlag London, Ltd., London; EDP Sciences, Les Ulis, 2011, DOI 10.1007/978-0-85729-073-1. MR2766419

[41] S. G. Dani and John Smillie, *Uniform distribution of horocycle orbits for Fuchsian groups*, Duke Math. J. **51** (1984), no. 1, 185–194, DOI 10.1215/S0012-7094-84-05110-X. MR744294

[42] Diana Davis, *Billiards, Surfaces and Geometry: a problem-centered approach*, https://dianadavis.github.io/billiards-book.pdf.

[43] Diana Davis and Samuel Lelièvre, *Periodic paths on the pentagon, double pentagon and golden L.*, Preprint, arXiv:1810.11310, 2018.

[44] Vincent Delecroix, Élise Goujard, Peter Zograf, and Anton Zorich, *Contribution of one-cylinder square-tiled surfaces to Masur-Veech volumes* (English, with English and French summaries), with an appendix by Philip Engel; Some aspects of the theory of dynamical systems: a tribute to Jean-Christophe Yoccoz. Vol. I, Astérisque **415** (2020), 223–274, DOI 10.24033/ast. MR4142454

[45] Vincent Delecroix, Pascal Hubert, and Samuel Lelièvre, *Diffusion for the periodic wind-tree model* (English, with English and French summaries), Ann. Sci. Éc. Norm. Supér. (4) **47** (2014), no. 6, 1085–1110, DOI 10.24033/asens.2234. MR3297155

[46] Vincent Delecroix and Anton Zorich, *Cries and whispers in wind-tree forests*, What's next?—the mathematical legacy of William P. Thurston, Ann. of Math. Stud., vol. 205, Princeton Univ. Press, Princeton, NJ, 2020, pp. 83–115, DOI 10.2307/j.ctvthhdvv.8. MR4205637

[47] Manfredo P. do Carmo, *Differential geometry of curves & surfaces*, revised & updated second edition of [MR0394451], Dover Publications, Inc., Mineola, NY, 2016. MR3837152

[48] A. Douady and J. Hubbard, *On the density of Strebel differentials*, Invent. Math. **30** (1975), no. 2, 175–179, DOI 10.1007/BF01425507. MR396936

[49] Benjamin Dozier, *Measure bound for translation surfaces with short saddle connections*, Geom. Funct. Anal. **33** (2023), no. 2, 421–467, DOI 10.1007/s00039-023-00636-9. MR4578463

[50] Paul Ehrenfest and Tatiana Ehrenfest, *The conceptual foundations of the statistical approach in mechanics*, translated by M. J. Moravcsik, Cornell University Press, Ithaca, NY, 1959. MR106571

[51] Charles Ehresmann, *Structures locales et structures infinitésimales* (French), C. R. Acad. Sci. Paris **234** (1952), 587–589. MR46736

[52] Manfred Einsiedler and Thomas Ward, *Ergodic theory with a view towards number theory*, Graduate Texts in Mathematics, vol. 259, Springer-Verlag London, Ltd., London, 2011, DOI 10.1007/978-0-85729-021-2. MR2723325

[53] Alex Eskin, *Counting problems in moduli space*, Handbook of dynamical systems. Vol. 1B, Elsevier B. V., Amsterdam, 2006, pp. 581–595, DOI 10.1016/S1874-575X(06)80034-2. MR2186249

[54] Alex Eskin, Maxim Kontsevich, and Anton Zorich, *Sum of Lyapunov exponents of the Hodge bundle with respect to the Teichmüller geodesic flow*, Publ. Math. Inst. Hautes Études Sci. **120** (2014), 207–333, DOI 10.1007/s10240-013-0060-3. MR3270590

[55] Alex Eskin, Gregory Margulis, and Shahar Mozes, *Upper bounds and asymptotics in a quantitative version of the Oppenheim conjecture*, Ann. of Math. (2) **147** (1998), no. 1, 93–141, DOI 10.2307/120984. MR1609447

[56] Alex Eskin, Jens Marklof, and Dave Witte Morris, *Unipotent flows on the space of branched covers of Veech surfaces*, Ergodic Theory Dynam. Systems **26** (2006), no. 1, 129–162, DOI 10.1017/S0143385705000234. MR2201941

[57] Alex Eskin and Howard Masur, *Asymptotic formulas on flat surfaces*, Ergodic Theory Dynam. Systems **21** (2001), no. 2, 443–478, DOI 10.1017/S0143385701001225. MR1827113

[58] Alex Eskin, Howard Masur, and Anton Zorich, *Moduli spaces of abelian differentials: the principal boundary, counting problems, and the Siegel-Veech constants*, Publ. Math. Inst. Hautes Études Sci. **97** (2003), 61–179, DOI 10.1007/s10240-003-0015-1. MR2010740

[59] Alex Eskin and Andrei Okounkov, *Asymptotics of numbers of branched coverings of a torus and volumes of moduli spaces of holomorphic differentials*, Invent. Math. **145** (2001), no. 1, 59–103, DOI 10.1007/s002220100142. MR1839286

[60] Alex Eskin and Curt McMullen, *Mixing, counting, and equidistribution in Lie groups*, Duke Math. J. **71** (1993), no. 1, 181–209, DOI 10.1215/S0012-7094-93-07108-6. MR1230290

[61] Alex Eskin and Maryam Mirzakhani, *Invariant and stationary measures for the* SL(2, ℝ) *action on moduli space*, Publ. Math. Inst. Hautes Études Sci. **127** (2018), 95–324, DOI 10.1007/s10240-018-0099-2. MR3814652

[62] Alex Eskin, Maryam Mirzakhani, and Amir Mohammadi, *Isolation, equidistribution, and orbit closures for the* SL(2, ℝ) *action on moduli space*, Ann. of Math. (2) **182** (2015), no. 2, 673–721, DOI 10.4007/annals.2015.182.2.7. MR3418528

[63] Kenneth Falconer, *Fractal geometry: Mathematical foundations and applications*, 2nd ed., John Wiley & Sons, Inc., Hoboken, NJ, 2003, DOI 10.1002/0470013850. MR2118797

[64] Benson Farb and Dan Margalit, *A primer on mapping class groups*, Princeton Mathematical Series, vol. 49, Princeton University Press, Princeton, NJ, 2012. MR2850125

[65] H. M. Farkas and I. Kra, *Riemann surfaces*, 2nd ed., Graduate Texts in Mathematics, vol. 71, Springer-Verlag, New York, 1992, DOI 10.1007/978-1-4612-2034-3. MR1139765

[66] William Feller, *An introduction to probability theory and its application*, John Wiley & Sons, 1961.

[67] Simion Filip, *Zero Lyapunov exponents and monodromy of the Kontsevich-Zorich cocycle*, Duke Math. J. **166** (2017), no. 4, 657–706, DOI 10.1215/00127094-3715806. MR3619303

[68] Simion Filip, *Notes on the multiplicative ergodic theorem*, Ergodic Theory Dynam. Systems **39** (2019), no. 5, 1153–1189, DOI 10.1017/etds.2017.68. MR3928611

[69] Giovanni Forni, *Deviation of ergodic averages for area-preserving flows on surfaces of higher genus*, Ann. of Math. (2) **155** (2002), no. 1, 1–103, DOI 10.2307/3062150. MR1888794

[70] Giovanni Forni, *Effective unique ergodicity and weak mixing of translation flows*, Preprint, arXiv:2311.02714, 2023.

[71] Giovanni Forni and Carlos Matheus, *Introduction to Teichmüller theory and its applications to dynamics of interval exchange transformations, flows on surfaces and billiards*, J. Mod. Dyn. **8** (2014), no. 3-4, 271–436, DOI 10.3934/jmd.2014.8.271. MR3345837

[72] Giovanni Forni, Carlos Matheus, and Anton Zorich, *Zero Lyapunov exponents of the Hodge bundle*, Comment. Math. Helv. **89** (2014), no. 2, 489–535, DOI 10.4171/CMH/325. MR3225454

[73] Ralph H. Fox and Richard B. Kershner, *Concerning the transitive properties of geodesics on a rational polyhedron*, Duke Math. J. **2** (1936), no. 1, 147–150, DOI 10.1215/S0012-7094-36-00213-2. MR1545913

[74] Alex Furman, *Random walks on groups and random transformations*, Handbook of dynamical systems, Vol. 1A, North-Holland, Amsterdam, 2002, pp. 931–1014, DOI 10.1016/S1874-575X(02)80014-5. MR1928529

[75] Harry Furstenberg, *The unique ergodicity of the horocycle flow*, Recent advances in topological dynamics (Proc. Conf. Topological Dynamics, Yale Univ., New Haven, Conn., 1972; in honor of Gustav Arnold Hedlund), Lecture Notes in Math., Vol. 318, Springer, Berlin-New York, 1973, pp. 95–115. MR393339

[76] H. Furstenberg and H. Kesten, *Products of random matrices*, Ann. Math. Statist. **31** (1960), 457–469, DOI 10.1214/aoms/1177705909. MR121828

[77] Frederick P. Gardiner, *Teichmüller theory and quadratic differentials*, Pure and Applied Mathematics (New York), A Wiley-Interscience Publication, John Wiley & Sons, Inc., New York, 1987. MR903027

[78] Ilya Gekhtman and Alex Wright, *Smillie's Theorem on closed SL(2, ℝ)-orbits of quadratic differentials*, accessed October 10, 2023, available at http://www-personal.umich.edu/~alexmw/Smillie.pdf.

[79] William M. Goldman, *Geometric structures on manifolds*, Graduate Studies in Mathematics, vol. 227, American Mathematical Society, Providence, RI, 2022, DOI 10.1090/gsm/227. MR4500072

[80] Elise Goujard, *Volumes of strata of moduli spaces of quadratic differentials: getting explicit values* (English, with English and French summaries), Ann. Inst. Fourier (Grenoble) **66** (2016), no. 6, 2203–2251. MR3580171

[81] Eugene Gutkin and Chris Judge, *Affine mappings of translation surfaces: geometry and arithmetic*, Duke Math. J. **103** (2000), no. 2, 191–213, DOI 10.1215/S0012-7094-00-10321-3. MR1760625

[82] Heini Halberstam and Klaus Friedrich Roth, *Sequences*, 2nd ed., Springer-Verlag, New York-Berlin, 1983. MR687978

[83] John L. Harer, *The virtual cohomological dimension of the mapping class group of an orientable surface*, Invent. Math. **84** (1986), no. 1, 157–176, DOI 10.1007/BF01388737. MR830043

[84] Michael-Robert Herman, *Sur la conjugaison différentiable des difféomorphismes du cercle à des rotations* (French), Inst. Hautes Études Sci. Publ. Math. **49** (1979), 5–233. MR538680

[85] W. Patrick Hooper, *Grid graphs and lattice surfaces*, Int. Math. Res. Not. IMRN **12** (2013), 2657–2698, DOI 10.1093/imrn/rns124. MR3071661

[86] Eberhard Hopf, *Fuchsian groups and ergodic theory*, Trans. Amer. Math. Soc. **39** (1936), no. 2, 299–314, DOI 10.2307/1989750. MR1501848

[87] Roger E. Howe and Calvin C. Moore, *Asymptotic properties of unitary representations*, J. Functional Analysis **32** (1979), no. 1, 72–96, DOI 10.1016/0022-1236(79)90078-8. MR533220

[88] John Hubbard and Howard Masur, *Quadratic differentials and foliations*, Acta Math. **142** (1979), no. 3-4, 221–274, DOI 10.1007/BF02395062. MR523212

[89] John Hamal Hubbard, *Teichmüller theory and applications to geometry, topology, and dynamics. Vol. 1, Teichmüller theory*, with contributions by Adrien Douady, William Dunbar, Roland Roeder, Sylvain Bonnot, David Brown, Allen Hatcher, Chris Hruska and Sudeb Mitra; with forewords by William Thurston and Clifford Earle, Matrix Editions, Ithaca, NY, 2006. MR2245223

[90] Pascal Hubert and Erwan Lanneau, *Veech groups without parabolic elements*, Duke Math. J. **133** (2006), no. 2, 335–346, DOI 10.1215/S0012-7094-06-13326-4. MR2225696

[91] Pascal Hubert, Erwan Lanneau, and Martin Möller, *The Arnoux-Yoccoz Teichmüller disc*, Geom. Funct. Anal. **18** (2009), no. 6, 1988–2016, DOI 10.1007/s00039-009-0706-y. MR2491696

[92] Pascal Hubert and Thomas A. Schmidt, *An introduction to Veech surfaces*, Handbook of dynamical systems. Vol. 1B, Elsevier B. V., Amsterdam, 2006, pp. 501–526, DOI 10.1016/S1874-575X(06)80031-7. MR2186246

[93] Pascal Hubert and Thomas A. Schmidt, *Infinitely generated Veech groups*, Duke Math. J. **123** (2004), no. 1, 49–69, DOI 10.1215/S0012-7094-04-12312-8. MR2060022

[94] A. Hurwitz, *Ueber algebraische Gebilde mit eindeutigen Transformationen in sich* (German), Math. Ann. **41** (1892), no. 3, 403–442, DOI 10.1007/BF01443420. MR1510753

[95] Jürgen Jost, *Compact Riemann surfaces: An introduction to contemporary mathematics*, 3rd ed., Universitext, Springer-Verlag, Berlin, 2006, DOI 10.1007/978-3-540-33067-7. MR2247485

[96] Anatole Katok, *Invariant measures of flows on orientable surfaces* (Russian), Dokl. Akad. Nauk SSSR **211** (1973), 775–778; English transl., SovietMath.Dokl. **14** (1973), 1104–1108. MR0331438

[97] Anatole Katok, *Interval exchange transformations and some special flows are not mixing*, Israel J. Math. **35** (1980), no. 4, 301–310, DOI 10.1007/BF02760655. MR594335

[98] Anatole Katok and Boris Hasselblatt, *Introduction to the modern theory of dynamical systems*, with a supplementary chapter by Katok and Leonardo Mendoza, Encyclopedia of Mathematics and its Applications, vol. 54, Cambridge University Press, Cambridge, 1995, DOI 10.1017/CBO9780511809187. MR1326374

[99] A. N. Zemljakov and A. B. Katok, *Topological transitivity of billiards in polygons* (Russian), Mat. Zametki **18** (1975), no. 2, 291–300. MR399423

[100] Svetlana Katok, *Fuchsian groups*, Chicago Lectures in Mathematics, University of Chicago Press, Chicago, IL, 1992. MR1177168

[101] Michael Keane, *Non-ergodic interval exchange transformations*, Israel J. Math. **26** (1977), no. 2, 188–196, DOI 10.1007/BF03007668. MR435353

[102] Richard Kenyon and John Smillie, *Billiards on rational-angled triangles*, Comment. Math. Helv. **75** (2000), no. 1, 65–108, DOI 10.1007/s000140050113. MR1760496

[103] Steven Kerckhoff, Howard Masur, and John Smillie, *Ergodicity of billiard flows and quadratic differentials*, Ann. of Math. (2) **124** (1986), no. 2, 293–311, DOI 10.2307/1971280. MR855297

[104] Harvey B. Keynes and Dan Newton, *A "minimal", non-uniquely ergodic interval exchange transformation*, Math. Z. **148** (1976), no. 2, 101–105, DOI 10.1007/BF01214699. MR409766

[105] A. Ya. Khinchin, *Continued fractions*, translated from the third (1961) Russian edition, with a preface by B. V. Gnedenko; reprint of the 1964 translation, Dover Publications, Inc., Mineola, NY, 1997. MR1451873

[106] J. F. C. Kingman, *The ergodic theory of subadditive stochastic processes*, J. Roy. Statist. Soc. Ser. B **30** (1968), 499–510. MR254907

[107] Maxim Kontsevich and Anton Zorich, *Connected components of the moduli spaces of Abelian differentials with prescribed singularities*, Invent. Math. **153** (2003), no. 3, 631–678, DOI 10.1007/s00222-003-0303-x. MR2000471

[108] Maxim Kontsevich and Anton Zorich, *Lyapunov exponents and Hodge theory,* 1997, available at https://arxiv.org/abs/hep-th/9701164.

[109] M. Kontsevich, *Lyapunov exponents and Hodge theory*, The mathematical beauty of physics (Saclay, 1996), Adv. Ser. Math. Phys., vol. 24, World Sci. Publ., River Edge, NJ, 1997, pp. 318–332. MR1490861

[110] Steven Lalley, *Kingman's subadditive ergodic theorem lecture notes,* accessed on October 10, 2023, available at http://galton.uchicago.edu/~lalley/Courses/Graz/Kingman.pdf.

[111] Serge Lang, $SL_2(\mathbf{R})$, Graduate Texts in Mathematics, vol. 105, reprint of the 1975 edition, Springer-Verlag, New York, 1985. MR803508

[112] Erwan Lanneau, *Connected components of the strata of the moduli spaces of quadratic differentials* (English, with English and French summaries), Ann. Sci. Éc. Norm. Supér. (4) **41** (2008), no. 1, 1–56, DOI 10.24033/asens.2062. MR2423309

[113] François Ledrappier, *Distribution des orbites des réseaux sur le plan réel* (French, with English and French summaries), C. R. Acad. Sci. Paris Sér. I Math. **329** (1999), no. 1, 61–64, DOI 10.1016/S0764-4442(99)80462-5. MR1703338

[114] K. Mahler, *On lattice points in n-dimensional star bodies. I. Existence theorems*, Proc. Roy. Soc. London Ser. A **187** (1946), 151–187, DOI 10.1098/rspa.1946.0072. MR17753

[115] Daniel Massart, *A short introduction to translation surfaces, Veech surfaces, and Teichmüller dynamics*, Surveys in geometry I, Springer, Cham, 2022, pp. 343–388, DOI 10.1007/978-3-030-86695-2_9. MR4404400

[116] Howard Masur, *Uniquely ergodic quadratic differentials*, Comment. Math. Helv. **55** (1980), no. 2, 255–266, DOI 10.1007/BF02566685. MR576605

[117] Howard Masur, *The growth rate of trajectories of a quadratic differential*, Ergodic Theory Dynam. Systems **10** (1990), no. 1, 151–176, DOI 10.1017/S0143385700005459. MR1053805

[118] Howard Masur, *Interval exchange transformations and measured foliations*, Ann. of Math. (2) **115** (1982), no. 1, 169–200, DOI 10.2307/1971341. MR644018

[119] Howard Masur, *Hausdorff dimension of the set of nonergodic foliations of a quadratic differential*, Duke Math. J. **66** (1992), no. 3, 387–442, DOI 10.1215/S0012-7094-92-06613-0. MR1167101

[120] Howard Masur, *Ergodic theory of translation surfaces*, Handbook of dynamical systems. Vol. 1B, Elsevier B. V., Amsterdam, 2006, pp. 527–547, DOI 10.1016/S1874-575X(06)80032-9. MR2186247

[121] Howard Masur and John Smillie, *Hausdorff dimension of sets of nonergodic measured foliations*, Ann. of Math. (2) **134** (1991), no. 3, 455–543, DOI 10.2307/2944356. MR1135877

[122] Carlos Matheus Santos, *Three lectures on square-tiled surfaces*, Panoramas et Synthèses **58** (2022), 77-99, https://if-summer2018.sciencesconf.org/data/pages/origamis_Grenoble_matheus_3.pdf.

[123] Curtis T. McMullen, *Billiards and Teichmüller curves on Hilbert modular surfaces*, J. Amer. Math. Soc. **16** (2003), no. 4, 857–885, DOI 10.1090/S0894-0347-03-00432-6. MR1992827

[124] Curtis T. McMullen, *Prym varieties and Teichmüller curves*, Duke Math. J. **133** (2006), no. 3, 569–590, DOI 10.1215/S0012-7094-06-13335-5. MR2228463

[125] Curtis T. McMullen, *Dynamics of $SL_2(\mathbb{R})$ over moduli space in genus two*, Ann. of Math. (2) **165** (2007), no. 2, 397–456, DOI 10.4007/annals.2007.165.397. MR2299738

[126] Curtis T. McMullen, *Teichmüller curves in genus two: discriminant and spin*, Math. Ann. **333** (2005), no. 1, 87–130, DOI 10.1007/s00208-005-0666-y. MR2169830

[127] Curtis T. McMullen, *Billiards, heights, and the arithmetic of non-arithmetic groups*, Invent. Math. **228** (2022), no. 3, 1309–1351, DOI 10.1007/s00222-022-01101-4. MR4419633

[128] Curtis T. McMullen, *Billiards and Teichmüller curves*, Bull. Amer. Math. Soc. (N.S.) **60** (2023), no. 2, 195–250, DOI 10.1090/bull/1782. MR4557380

[129] Yair Minsky and Barak Weiss, *Nondivergence of horocyclic flows on moduli space*, J. Reine Angew. Math. **552** (2002), 131–177, DOI 10.1515/crll.2002.088. MR1940435

[130] Martin Möller, *Shimura and Teichmüller curves*, J. Mod. Dyn. **5** (2011), no. 1, 1–32, DOI 10.3934/jmd.2011.5.1. MR2787595

[131] Martin Möller, *Variations of Hodge structures of a Teichmüller curve*, J. Amer. Math. Soc. **19** (2006), no. 2, 327–344, DOI 10.1090/S0894-0347-05-00512-6. MR2188128

[132] Thierry Monteil, *Introduction to the theorem of Kerkhoff, Masur and Smillie*, in Arbeitsgemeinschaft: Mathematical Billards, Abstracts from the session held April 4–10, 2010; organized by Sergei Tabachnikov and Serge Troubetzkoy; Oberwolfach Reports. Vol. 7, no. 2, Oberwolfach Rep. **7** (2010), no. 2, 989–994, DOI 10.4171/OWR/2010/17. MR2768157

[133] Dave Witte Morris, *Ratner's theorems on unipotent flows*, Chicago Lectures in Mathematics, University of Chicago Press, Chicago, IL, 2005. MR2158954

[134] Lee Mosher, *Tiling the projective foliation space of a punctured surface*, Trans. Amer. Math. Soc. **306** (1988), no. 1, 1–70, DOI 10.2307/2000830. MR927683

[135] Amos Nevo, *Equidistribution in measure-preserving actions of semisimple groups: case of $SL_2(\mathbb{R})$*, Preprint, arXiv:1708.03886, 2017.

[136] Amos Nevo, Rene Rühr, and Barak Weiss, *Effective counting on translation surfaces*, Adv. Math. **360** (2020), 106890, 29, DOI 10.1016/j.aim.2019.106890. MR4031118

[137] Jakob Nielsen, *A theorem on the topology of surface transformations* (Danish), Norsk Mat. Tidsskr. **23** (1941), 5. MR13307

[138] V. I. Oseledec, *A multiplicative ergodic theorem. Characteristic Ljapunov, exponents of dynamical systems* (Russian), Trudy Moskov. Mat. Obšč. **19** (1968), 179–210. MR240280

[139] Jean-François Quint, *Rigidité des $SL_2(\mathbb{R})$-orbites dans les espaces de modules de surfaces plates [d'après Eskin, Mirzakhani et Mohammadi]* (French), Astérisque **380**, **Séminaire Bourbaki. Vol. 2014/2015** (2016), Exp. No. 1092, 83–138. MR3522172

[140] Marina Ratner, *Interactions between ergodic theory, Lie groups, and number theory*, Proceedings of the International Congress of Mathematicians, Vol. 1, 2 (Zürich, 1994), Birkhäuser, Basel, 1995, pp. 157–182. MR1403920

[141] H. L. Royden, *Automorphisms and isometries of Teichmüller space*, Advances in the Theory of Riemann Surfaces (Proc. Conf., Stony Brook, N.Y., 1969), Ann. of Math. Stud., No. 66, Princeton Univ. Press, Princeton, NJ, 1971, pp. 369–383. MR288254

[142] H. L. Royden, *Real analysis*, 3rd ed., Macmillan Publishing Company, New York, 1988. MR1013117

[143] Walter Rudin, *Principles of mathematical analysis*, 3rd ed., International Series in Pure and Applied Mathematics, McGraw-Hill Book Co., New York-Auckland-Düsseldorf, 1976. MR385023

[144] Walter Rudin, *Real and complex analysis*, 3rd ed., McGraw-Hill Book Co., New York, 1987. MR924157

[145] Walter Rudin, *Functional analysis*, 2nd ed., International Series in Pure and Applied Mathematics, McGraw-Hill, Inc., New York, 1991. MR1157815

[146] E. A. Sataev, *The number of invariant measures for flows on orientable surfaces* (Russian), Izv. Akad. Nauk SSSR Ser. Mat. **39** (1975), no. 4, 860–878. MR391184

[147] Wilhelm Schlag, *A course in complex analysis and Riemann surfaces*, Graduate Studies in Mathematics, vol. 154, American Mathematical Society, Providence, RI, 2014, DOI 10.1090/gsm/154. MR3186310

[148] Thomas A. Schmidt and Mark Sheingorn, *Covering the Hecke triangle surfaces*, Ramanujan J. **1** (1997), no. 2, 155–163, DOI 10.1023/A:1009716101565. MR1606184

[149] Gabriela Schmithüsen, *Examples for Veech groups of origamis*, The geometry of Riemann surfaces and abelian varieties, Contemp. Math., vol. 397, Amer. Math. Soc., Providence, RI, 2006, pp. 193–206, DOI 10.1090/conm/397/07473. MR2218009

[150] Richard Evan Schwartz, *Mostly surfaces*, Student Mathematical Library, vol. 60, American Mathematical Society, Providence, RI, 2011, DOI 10.1090/stml/060. MR2809109

[151] Sol Schwartzman, *Asymptotic cycles*, Ann. of Math. (2) **66** (1957), 270–284, DOI 10.2307/1969999. MR88720

[152] Carl Ludwig Siegel, *A mean value theorem in geometry of numbers*, Ann. of Math. (2) **46** (1945), 340–347, DOI 10.2307/1969027. MR12093

[153] Carl Ludwig Siegel, *Lectures on the geometry of numbers*, notes by B. Friedman; rewritten by Komaravolu Chandrasekharan with the assistance of Rudolf Suter; with a preface by Chandrasekharan, Springer-Verlag, Berlin, 1989, DOI 10.1007/978-3-662-08287-4. MR1020761

[154] John Smillie and Corinna Ulcigrai, *Beyond Sturmian sequences: coding linear trajectories in the regular octagon*, Proc. Lond. Math. Soc. (3) **102** (2011), no. 2, 291–340, DOI 10.1112/plms/pdq018. MR2769116

[155] John Smillie and Barak Weiss, *Characterizations of lattice surfaces*, Invent. Math. **180** (2010), no. 3, 535–557, DOI 10.1007/s00222-010-0236-0. MR2609249

[156] John Smillie and Barak Weiss, *Veech's dichotomy and the lattice property*, Ergodic Theory Dynam. Systems **28** (2008), no. 6, 1959–1972, DOI 10.1017/S0143385708000114. MR2465608

[157] Kurt Strebel, *Quadratic differentials*, Ergebnisse der Mathematik und ihrer Grenzgebiete (3) [Results in Mathematics and Related Areas (3)], vol. 5, Springer-Verlag, Berlin, 1984, DOI 10.1007/978-3-662-02414-0. MR743423

[158] Diaaeldin Taha, *On Cross Sections to the Horocycle and Geodesic Flows on Quotients by Hecke Triangle Groups*, ProQuest LLC, Ann Arbor, MI, 2019. Thesis (Ph.D.)–University of Washington. MR4082754

[159] William P. Thurston, *On the geometry and dynamics of diffeomorphisms of surfaces*, Bull. Amer. Math. Soc. (N.S.) **19** (1988), no. 2, 417–431, DOI 10.1090/S0273-0979-1988-15685-6. MR956596

[160] Caglar Uyanik and Grace Work, *The distribution of gaps for saddle connections on the octagon*, Int. Math. Res. Not. IMRN **18** (2016), 5569–5602, DOI 10.1093/imrn/rnv317. MR3567252

[161] William A. Veech, *Strict ergodicity in zero dimensional dynamical systems and the Kronecker-Weyl theorem* mod 2, Trans. Amer. Math. Soc. **140** (1969), 1–33, DOI 10.2307/1995120. MR240056

[162] William A. Veech, *Gauss measures for transformations on the space of interval exchange maps*, Ann. of Math. (2) **115** (1982), no. 1, 201–242, DOI 10.2307/1971391. MR644019

[163] William A. Veech, *The Teichmüller geodesic flow*, Ann. of Math. (2) **124** (1986), no. 3, 441–530, DOI 10.2307/2007091. MR866707

[164] W. A. Veech, *Teichmüller curves in moduli space, Eisenstein series and an application to triangular billiards*, Invent. Math. **97** (1989), no. 3, 553–583, DOI 10.1007/BF01388890. MR1005006

[165] William A. Veech, *Siegel measures*, Ann. of Math. (2) **148** (1998), no. 3, 895–944, DOI 10.2307/121033. MR1670061

[166] Marcelo Viana, *Lyapunov exponents of Teichmüller flows*, Partially hyperbolic dynamics, laminations, and Teichmüller flow, Fields Inst. Commun., vol. 51, Amer. Math. Soc., Providence, RI, 2007, pp. 139–201. MR2388695

[167] Marcelo Viana, *Ergodic theory of interval exchange maps*, Rev. Mat. Complut. **19** (2006), no. 1, 7–100, DOI 10.5209/rev_REMA.2006.v19.n1.16621. MR2219821

[168] Ya. B. Vorobets, *Plane structures and billiards in rational polygons: the Veech alternative* (Russian), Uspekhi Mat. Nauk **51** (1996), no. 5(311), 3–42, DOI 10.1070/RM1996v051n05ABEH002993; English transl., Russian Math. Surveys **51** (1996), no. 5, 779–817. MR1436653

[169] Clayton C. Ward, *Calculation of Fuchsian groups associated to billiards in a rational triangle*, Ergodic Theory Dynam. Systems **18** (1998), no. 4, 1019–1042, DOI 10.1017/S0143385798117479. MR1645350

[170] Peter Walters, *An introduction to ergodic theory*, Graduate Texts in Mathematics, vol. 79, Springer-Verlag, New York-Berlin, 1982. MR648108

[171] Hermann Weyl, *Über die Gleichverteilung von Zahlen mod. Eins* (German), Math. Ann. **77** (1916), no. 3, 313–352, DOI 10.1007/BF01475864. MR1511862

[172] Alex Wright, *Schwarz triangle mappings and Teichmüller curves: the Veech-Ward-Bouw-Möller curves*, Geom. Funct. Anal. **23** (2013), no. 2, 776–809, DOI 10.1007/s00039-013-0221-z. MR3053761

[173] Alex Wright, *From rational billiards to dynamics on moduli spaces*, Bull. Amer. Math. Soc. (N.S.) **53** (2016), no. 1, 41–56, DOI 10.1090/bull/1513. MR3403080

[174] Jean-Christophe Yoccoz, *Interval exchange maps and translation surfaces*, Homogeneous flows, moduli spaces and arithmetic, Clay Math. Proc., vol. 10, Amer. Math. Soc., Providence, RI, 2010, pp. 1–69. MR2648692

[175] Anton Zorich, *Asymptotic flag of an orientable measured foliation on a surface*, Geometric study of foliations (Tokyo, 1993), World Sci. Publ., River Edge, NJ, 1994, pp. 479–498. MR1363744

[176] Anton Zorich, *How do the leaves of a closed 1-form wind around a surface?*, Pseudoperiodic topology, Amer. Math. Soc. Transl. Ser. 2, vol. 197, Amer. Math. Soc., Providence, RI, 1999, pp. 135–178, DOI 10.1090/trans2/197/05. MR1733872

[177] Anton Zorich, *Flat surfaces*, Frontiers in number theory, physics, and geometry. I, Springer, Berlin, 2006, pp. 437–583. MR2261104

[178] Anton Zorich, *Le théorème de la baguette magique de A. Eskin et M. Mirzakhani* (French), Gaz. Math. **142** (2014), 39–54. MR3278429

Index

$GL^+(2,\mathbb{R})$-action, 3, 39, 48

Abelian differential, 21, 33
Affine diffeomorphisms, 142
Affine group, 159
Arnoux-Yoccoz surfaces, 149–150
Asymptotic flag, 111

Birkhoff Ergodic Theorem, 68
Boundaries, 139
Branched cover, 149

Canonical bundle, 35
Choquet simplex, 82
Components of general strata, 50–51
Cone angle, *see also* Singularity, 10, 15, 34
Continued fractions, 72, 117
Counting
   and volumes, 162
   in orbits, 156
   lattice points, 123
   saddle connections, 127, 135
Cylinder, 26

Dehn twists, 40
Delaunay decomposition, 53–54
Delaunay triangulation, 31, 52, 56
Divisors, 22

Eierlegende Wollmichsau, 161–162
Equidistribution, 4–5, 132, 135
Ergodic, 63, 67–68, 81–82
   Birkhoff Ergodic Theorem, 68
   Kingman's subadditive ergodic theorem, 114
   Nevo's ergodic theorem, 132, 135
   nonergodicity, 74, 119
   nonunique, 119–120
   Oseledets's theorem, 115
   unique ergodicity, 69
Eskin-Mirzakhani-Mohammadi theorems, 108
Euler characteristic, 11
Exponential mixing, 108–109, 136

First return, 4, 84, 87

Gauss-Bonnet formula, 20
Generic, 69
Generic point, 69, 97

Haar measure, 124
Half-translation surface, 33–35
Hausdorff dimension, 81, 119
Hecke triangle group, 146
Higher-order differential, 36, 50
Holomorphic 1-form, *see also* Abelian differential, 1, 10, 21
Holonomy vector, 24–26, 28, 127
Hopf argument, 100
Horocycle flow, 107–108
Howe-Moore theorem, 107
Hyperelliptic strata, 89

Interval exchange transformations, 63, 86–87
Invariant measure, 87–88, 96–97

Jacobian, 159–160

Kingman's ergodic theorem, 114–115
Kontsevich-Zorich cocycle, 116–117

L-shape surfaces, 145
Lagrangian subspaces, 84
Lattice surface, 71, 124, 141, 144
    square-tiled surface, *see also* Origami, 160
Lyapunov exponent, 112–113

Mapping class group, 40, 43
Marking, 41
Masur's criterion, 96
Masur-Smillie-Veech measure, 46–47, 59–61
McMullen's theorem, 14
Measure preserving, 67–68
    flow, 68
    transformation, 68
Measured foliation, 100–101
Meromorphic differential, 36
    higher-order differential, 36, 50
Minimal direction, 66
Minimal domain, 64–65
Mixing, 89–93
    exponential, 108, 136
    quantitative weak, 118
    strong, 90, 91
    weak, 90–91, 117, 118
Moduli space, 39
MSV measure, 59

Nevo's theorem, 132, 135
No small triangles, 157–158
Nonergodicity, 74

Orbit
    closed, 156
    closure, 107
    counting, 156
Origami, 144
Oseledets's theorem, 115–116

Period coordinates, 45–46
Periodic orbit, 5
Platonic solid, 37
Poincaré recurrence, 99–100
Pseudo-Anosov, 40

Quadratic differential, *see also*
    Half-translation surface, 33–35
Quantitative weak mixing, 118

Quasiconformal map, 41
Quasimodular form, 163
Quasimodularity, 163

Rational billiards, 16–18
Reducible, 40
Riemann surface, 1, 10
Riemann-Roch theorem, 22
Royden's theorem, 41

Saddle connection, 26–27, 127
Schwartzmann asymptotic cycle, 85
Separatrices, 25
Siegel's formula, 125–126
Siegel-Veech
    constant, 128
    formula, 128
    transform, 128, 133
Singularity, 82
Skew-product, 71
Slit construction, 14
Smillie's theorem, 153
Spectral gap, 109
Square-tiled, 144–145
Square-tiled surface, 145, 160–163
Stone-Weierstrass theorem, 6
Straight line flow, 10
Strata, 12, 44
    components of, 50
    hyperelliptic, 50, 89
Strebel direction, 66–67, 150–151, 155, 158
Strong mixing, 90, 91

Teichmüller curve, 153
Teichmüller flow, 95–96, 102
    geodesic, 49
    horocycle, 49
Teichmüller metric, 41
Teichmüller space, 39–40
Trace field, 159
Translation surface
    polygonal presentation, 9
Transverse measure, 82–83
Triangulation, 30
    Delaunay triangulation, 31, 52–56

Unfolding, 16–17
Unimodular lattice, 124
Unique ergodicity, 69

Veech dichotomy, 151
Veech group, 142–143, 159
Voronoi decomposition, 53

Index

Weak mixing, 90, 91, 117–118
Weyl's theorem, 6
Whitehead equivalence, 101
Wind-tree, 119

Zippered rectangles, 89, 101
Zorich phenomenon, 110

Selected Published Titles in This Series

242 **Jayadev S. Athreya and Howard Masur,** Translation Surfaces, 2024
238 **Julio González-Díaz, Ignacio García-Jurado, and M. Gloria Fiestras-Janeiro,** An Introductory Course on Mathematical Game Theory and Applications, Second Edition, 2023
237 **Michael Levitin, Dan Mangoubi, and Iosif Polterovich,** Topics in Spectral Geometry, 2023
235 **Bennett Chow,** Ricci Solitons in Low Dimensions, 2023
234 **Andrea Ferretti,** Homological Methods in Commutative Algebra, 2023
233 **Andrea Ferretti,** Commutative Algebra, 2023
232 **Harry Dym,** Linear Algebra in Action, Third Edition, 2023
231 **Luís Barreira and Yakov Pesin,** Introduction to Smooth Ergodic Theory, Second Edition, 2023
230 **Barbara Kaltenbacher and William Rundell,** Inverse Problems for Fractional Partial Differential Equations, 2023
229 **Giovanni Leoni,** A First Course in Fractional Sobolev Spaces, 2023
228 **Henk Bruin,** Topological and Ergodic Theory of Symbolic Dynamics, 2022
227 **William M. Goldman,** Geometric Structures on Manifolds, 2022
226 **Milivoje Lukić,** A First Course in Spectral Theory, 2022
225 **Jacob Bedrossian and Vlad Vicol,** The Mathematical Analysis of the Incompressible Euler and Navier-Stokes Equations, 2022
224 **Ben Krause,** Discrete Analogues in Harmonic Analysis, 2022
223 **Volodymyr Nekrashevych,** Groups and Topological Dynamics, 2022
222 **Michael Artin,** Algebraic Geometry, 2022
221 **David Damanik and Jake Fillman,** One-Dimensional Ergodic Schrödinger Operators, 2022
220 **Isaac Goldbring,** Ultrafilters Throughout Mathematics, 2022
219 **Michael Joswig,** Essentials of Tropical Combinatorics, 2021
218 **Riccardo Benedetti,** Lectures on Differential Topology, 2021
217 **Marius Crainic, Rui Loja Fernandes, and Ioan Mărcuţ,** Lectures on Poisson Geometry, 2021
216 **Brian Osserman,** A Concise Introduction to Algebraic Varieties, 2021
215 **Tai-Ping Liu,** Shock Waves, 2021
214 **Ioannis Karatzas and Constantinos Kardaras,** Portfolio Theory and Arbitrage, 2021
213 **Hung Vinh Tran,** Hamilton–Jacobi Equations, 2021
212 **Marcelo Viana and José M. Espinar,** Differential Equations, 2021
211 **Mateusz Michałek and Bernd Sturmfels,** Invitation to Nonlinear Algebra, 2021
210 **Bruce E. Sagan,** Combinatorics: The Art of Counting, 2020
209 **Jessica S. Purcell,** Hyperbolic Knot Theory, 2020
208 **Vicente Muñoz, Ángel González-Prieto, and Juan Ángel Rojo,** Geometry and Topology of Manifolds, 2020
207 **Dmitry N. Kozlov,** Organized Collapse: An Introduction to Discrete Morse Theory, 2020
206 **Ben Andrews, Bennett Chow, Christine Guenther, and Mat Langford,** Extrinsic Geometric Flows, 2020
205 **Mikhail Shubin,** Invitation to Partial Differential Equations, 2020
204 **Sarah J. Witherspoon,** Hochschild Cohomology for Algebras, 2019
203 **Dimitris Koukoulopoulos,** The Distribution of Prime Numbers, 2019

For a complete list of titles in this series, visit the
AMS Bookstore at **www.ams.org/bookstore/gsmseries/**.